Power Quality and Harmonics Management in Modern Power Systems

Edited by Muhyaddin Rawa, Ziad M. Ali and Shady H.E. Abdel Aleem

Published in London, United Kingdom

Power Quality and Harmonics Management in Modern Power Systems
http://dx.doi.org/10.5772/intechopen.111082
Edited by Muhyaddin Rawa, Ziad M. Ali and Shady H.E. Abdel Aleem

Contributors
Abdourahimoun Daouda, Alexandru Gabriel Gheorghe, Arvind Singh, Busireddy Hemanth Kumar, Chamberlin Stéphane Azebaze Mboving, Florin Constantinescu, Florin Roman Enache, Joseph John Puthenkalam, Mihai Eugen Marin, Muhyaddin Rawa, Nagaraj Ramrao, Ponguleti Sandhya, Rangu Seshu Kumar, Shady H.E. Abdel Aleem, Smart Edward Amanfo, Tomasz Siostrzonek, Zbigniew Hanzelka, Ziad M. Ali

First published in London, United Kingdom, 2024 by IntechOpen
IntechOpen is the global imprint of INTECHOPEN LIMITED, registered in England and Wales, registration number: 11086078, 167-169 Great Portland Street, London, W1W 5PF, United Kingdom

British Library Cataloguing-in-Publication Data
A catalogue record for this book is available from the British Library

Additional hard and PDF copies can be obtained from orders@intechopen.com

Power Quality and Harmonics Management in Modern Power Systems
Edited by Muhyaddin Rawa, Ziad M. Ali and Shady H.E. Abdel Aleem
p. cm.
Print ISBN 978-1-83769-817-2
Online ISBN 978-1-83769-816-5
eBook (PDF) ISBN 978-1-83769-818-9

For EU product safety concerns:
IN TECH d.o.o., Prolaz Marije Krucifikse Kozulić 3, 51000 Rijeka, Croatia,
info@intechopen.com or visit our website at intechopen.com.

Meet the editors

Muhyaddin Rawa received a Ph.D in electrical and electronic engineering from the University of Nottingham in 2014. He has more than seven years of experience in the Saudi Electricity Company. Currently, he is an associate professor with the Department of Electrical and Computer Engineering at King Abdulaziz University, where he is the deputy director of the Center of Research Excellence in Renewable Energy and Power Systems. He is actively involved in industrial consultancy for major corporations in power systems projects. His research interests include power quality, renewable energy, and smart grids. Dr. Muhyaddin has published 155+ journal and conference papers, holds 17 U.S. patents, and has authored five book chapters.

Ziad M. Ali received a BSc degree in electrical engineering from Assiut University, Faculty of Engineering, Assuit, Egypt, in 1998. He worked as a demonstrator in the Aswan Faculty of Engineering, South Valley University, Aswan, Egypt. He obtained an MSc degree in electrical engineering from Assiut University, Faculty of Engineering in 2003. He worked as an assistant lecturer at the Aswan Faculty of Engineering. He obtained a Ph.D. degree in 2010 from Kazan State Technical University, Tatarstan, Russia. He is currently working as an associate professor in the Electrical Department at the College of Engineering at Wadi Addawasir, Prince Sattam bin Abdulaziz University, Saudi Arabia. He is among the most influential scholars in the world according to the classification made by Stanford University (USA).

Shady H. E. Abdel Aleem (M'12, SM'21) received BSc, MSc, and Ph.D. degrees in electrical power and machines from the Faculty of Engineering, Helwan University, Egypt, in 2002, and the Faculty of Engineering, Cairo University, Egypt, in 2010 and 2013, respectively. Currently, he is an associate professor at the Department of Electrical Engineering, Institute of Aviation Engineering and Technology, Giza, Egypt. He is the vice dean for education and student affairs. Also, he is a consultant of power quality studies in the ETA Electric Company, Egypt. His research interests include harmonic problems in power systems, power quality, renewable energy, smart grid, energy efficiency, optimization, green energy, and economics. Dr. Shady is the author or co-author of many refereed journals and conference papers. He has published more than 180 journal and conference papers, more than 20 book chapters, and 10 edited books with the Institution of Engineering and Technology (IET), Elsevier, Springer, CRC Press, and InTech. He was awarded the State Encouragement Award in Engineering Sciences in 2017 from Egypt. He was also awarded the Medal of Distinction from the first class of the Egyptian State Award in 2020 from Egypt. Dr. Shady is a senior member of the Institute of Electrical and Electronics Engineers (IEEE). Dr. Shady is also a member of the Institution of Engineering and Technology (IET). He is an editor, guest editor, or associate editor for the *International*

Journal of Renewable Energy Technology, Vehicle Dynamics, IET Journal of Engineering, Energies, Sustainability, Technology and Economics of Smart Grids and Sustainable Energy, and others. He is among the most influential scholars in the world according to the classification made by Stanford University (USA).

Contents

Preface

This book aims to understand power quality and harmonics management comprehensively. It covers various topics, from the challenges of maintaining power quality in the presence of distributed generation and power electronic-based technologies to the impact of harmonics on electrical networks. The book explores frequency domain models of nonlinear loads, discusses the mitigation of power system harmonics through active filters, investigates the performance efficiency of shunt hybrid systems, examines the generation of higher harmonics in hoisting machinery in underground mines, and offers a multidisciplinary perspective on energy security in the face of climate change and uncertainty. Each chapter outlines the problem's fundamental structure to provide a basic understanding of the approaches discussed. It is structured and arranged into seven chapters.

Chapter 1, "Perspective Chapter: Power Quality and Hosting Capacity", discusses the challenges and importance of maintaining acceptable power quality (PQ) levels in the context of increasing distributed generation (DG) and power electronic-based technologies. The chapter overviews PQ definitions, disturbances, causes, and standards, focusing on harmonic issues. It explores the impact of nonlinear loads and novel load equipment on PQ and emphasizes the significance of PQ in a competitive energy environment. The chapter also discusses renewable-based DG and hosting capacity studies, including types, challenges, and solutions. Additionally, it presents a literature overview of existing solutions for harmonic management. The chapter highlights the need for a comprehensive approach to PQ that considers current and voltage quality and emphasizes the development of standards to address PQ issues stemming from nonlinear loads and renewable energy sources.

Chapter 2, "Impact of Harmonics on the Electrical Network Distribution", focuses on the impact of harmonics on electrical networks and discusses the consequences of harmonics generated by electronic devices on the electrical grid. It emphasizes the need to control and mitigate harmonics as they can cause premature deterioration of electrical loads and power supply devices. The chapter explores the distinction between linear and nonlinear loads, with linear loads drawing quasi-sinusoidal currents and nonlinear loads distorting the voltage waveforms. It also delves into fundamental indicators of electrical harmonics, including power factor, order of harmonics, effective value of harmonic quantities, and total harmonic distortion (THD). The importance of measuring and analyzing harmonics using electrical analyzers is highlighted.

Chapter 3, "Perspective Chapter: Frequency Domain Models of Nonlinear Loads in Power Systems", focuses on frequency domain models of nonlinear loads in power systems. It begins by discussing harmonic current source models for one- and two-diode rectifiers, with their parameters determined through simulations. The next section presents frequency domain models for nonlinear home appliances, with parameters determined through measurement. It also introduces frequency domain models for loads using firing angle control devices. The chapter then explores simulations of a large circuit in both the time domain and frequency domain using CADENCE and ADS algorithms. The harmonic balance analysis of ADS using the proposed frequency domain models is the most efficient. Finally, the chapter emphasizes the importance of understanding and modeling harmonics in power systems and the use of frequency domain analysis for studying nonlinear circuits.

Chapter 4, "Perspective Chapter: Mitigation of Power System Harmonics with the Incorporation of Active Filter for a Radial Distribution System", discusses the problem of harmonics in power systems and explores the effectiveness of active filters in mitigating harmonics. The research work evaluates the performance of active filters using the sparrow search optimization technique in a MATLAB environment. The simulation results, obtained using a standard IEEE 13 bus test system and an unbalanced power system, demonstrate a significant reduction in harmonics by incorporating active filters. The chapter highlights that hybrid active filters provide the best harmonic mitigation performance and emphasizes the economic feasibility of using active filters for reducing harmonics. It also emphasizes the importance of power quality in distributed power systems and the role of shunt active filters in mitigating harmonic distortion. The chapter further discusses PQ indices for assessing the impact of harmonics and the integration of active filters in distribution systems. Finally, the chapter provides valuable insights into applying active filters for power system harmonics mitigation and their potential to improve the quality and reliability of power systems.

Chapter 5, "Investigation on the Performance Efficiency of the Shunt Hybrid Active Power Filter", focuses on investigating the performance efficiency of the shunt hybrid system. The specific details and findings of the investigation are not provided in the truncated content. The chapter explores the operation and characteristics of the shunt hybrid system in power systems. It discusses this type of system's advantages, challenges, and performance factors. The investigation may include experimental data, simulations, or analytical models to evaluate the efficiency and effectiveness of the shunt hybrid system. The chapter aims to contribute to the understanding and optimization of shunt hybrid systems in power system applications.

Chapter 6, "The Hoisting Machines as Source of Higher Harmonics in Underground Mines", focuses on the structure and characteristics of drive systems used in hoisting machinery in underground mines. The chapter highlights the use of DC reciprocating motors and complex (multipulse) converters in these systems. Special attention is given to the impact of these systems on the power supply network, particularly the generation of higher harmonics. The chapter presents measurement results conducted under real conditions, both before and after installing new solutions to improve machine efficiency. The keywords associated with this chapter include power converters, hoisting machines, multipulse systems, power quality, DC drive systems, and power grids in underground mines.

Chapter 7, "Perspective Chapter: The Regime Matters – A Multidisciplinary Perspective on Energy Security in the Era of Climate Change and Growing Uncertainties for Resilience in Sustainable Energy Development", explores the concept of energy security in the context of climate change and increasing uncertainties. The chapter discusses various definitions of energy security and its positioning within the international political economy of energy. It examines energy security from multiple perspectives, including those of energy importing and exporting nations, transit regions, militarization, energy shocks, demographic shifts, and corruption. The chapter emphasizes energy security's challenges, risks, and vulnerabilities and highlights its intricate interconnections. It concludes by advocating for integrating resilience thinking into energy security policies due to the growing uncertainties in social, economic, and ecological systems compounded by climate change. The chapter also discusses energy security, climate change, uncertainty regimes, supply–demand security, resilience, and sustainable energy development.

The editors hope readers find this book valuable in gaining insights into the various aspects of power quality and harmonics management and the practical approaches and solutions employed in the field. Whether for researchers, engineers, or professionals working in the energy sector, this book provides a solid foundation for understanding and addressing the challenges and complexities associated with power quality and harmonics in modern power systems.

Muhyaddin Rawa
Smart Grids Research Group,
Center of Research Excellence in Renewable Energy and Power Systems,
King Abdulaziz University,
Jeddah, Saudi Arabia

Faculty of Engineering,
Department of Electrical and Computer Engineering,
K. A. CARE Energy Research and Innovation Center,
Jeddah, Saudi Arabia

Ziad M. Ali
Electrical Engineering Department,
College of Engineering at Wadi Addawaser,
Prince Sattam bin Abdulaziz University,
Wadi Addawaser, Saudi Arabia

Shady H.E. Abdel Aleem
Associate Professor,
Department of Electrical Engineering,
Institute of Aviation Engineering and Technology,
Giza, Egypt

Section 1

Harmonics Management in Modern Power Systems

Chapter 1

Perspective Chapter: Power Quality and Hosting Capacity

Muhyaddin Rawa, Ziad M. Ali and Shady H.E. Abdel Aleem

Abstract

With the increasing prevalence of distributed generation (DG) and power electronic-based technologies, consumers will have more alternatives for obtaining energy from different public or private sources. The issues will be with power quality (PQ), pricing, and reliability. Shortly, maintaining acceptable power quality levels above certain acceptable thresholds will be challenging because of the special difficulties brought on by nonlinear loads and novel types of load equipment. The significance of current and voltage quality issues increases even further in such an environment of competition. The chapter is dedicated to presenting an overview of PQ definitions, disturbances, causes, and standards. Harmonic description, sources, effects, and harmonic filtering techniques are also presented. Then, renewable-based DGs and HC studies—types, challenges, and solutions, are demonstrated. Further, a literature overview of the existing solutions under consideration (harmonic management) is presented and discussed.

Keywords: distributed generation, filters, harmonics, harmonics distortion, hosting capacity, optimization, power quality

1. Introduction

Improving electrical power quality is an intention agreed on by consumers and electrical utilities. The primary goals in terms of power quality (PQ) are generating clean power—that is, power that is not distorted—and cost-effectively delivering it to customers with adequate technical performance. Advancements in semiconductor technology have brought about harmonic pollution and other PQ concerns, a range of nonlinear load types, and the installation of renewable energy resources that rely on power electronic-based equipment (rectifiers and inverters) for their operations. It became clear that harmonics management is now an essential issue in power systems rather than a secondary concern [1, 2]. Power grid performance can be adversely affected by several factors, including overloading and increased heating of lines and cables, frequency-dependent equipment, diminished voltage quality, decreased transmission efficiency, an increase in energy losses during transmission and distribution, and deterioration of true and displacement power factors of loads [3].

Modern sophisticated converters with frequency-coupling dynamics also introduce harmonic instability (e.g., resonance, amplification of voltage or current, or unusual harmonics in the high-frequency spectrum) [4]. This means that PQ is so essential to contemporary energy systems.

As distributed generation (DGs) become more prevalent and utilize power electronic-based technologies, consumers will have more options when it comes to purchasing energy from various public or private sources. The cost, reliability, and quality of electricity will be the problems. Maintaining acceptable PQ levels above specific acceptable limits will be a major challenge in the upcoming decades due to the unique challenges presented by new types of load equipment and nonlinear loads. In such a competitive scenario, the importance of PQ issues becomes even more significant than before [5]. Furthermore, since they can now create and sell electricity through their DGs (prosumers), consumers connected to an electrical grid are no longer considered consumers in the context of the deregulated power market and many energy providers. Enhancing the PQ performance of the systems thus gets more difficult due to new and developing complications.

PQ provides several justifications to the various electric entities' stakeholders. Some perceive PQ as the voltage quality, others as the current quality (based on amperage), and others as the system's dependability. PQ, for example, is defined as "the powering and grounding concept of sensitive electronic equipment in a manner appropriate for the equipment" in IEEE Standard 1100 [6].

Put simply, each entity defines it according to its own perspective. It is possible to determine the cause and responsibility of a disturbance wrongly when a general term like PQ has an ambiguous definition. This was obvious in the surveys carried out by the Georgian Power Company [7, 8] for the causes of PQ disturbances. These surveys clearly showed how divergent the utility and customer perspectives are from one another. But both attribute two-thirds of the problems to lightning and other natural occurrences [9, 10]. This indicates that, under typical circumstances, PQ maintains a nearly pure sinusoidal waveform of voltages and currents. Generally speaking, deviations of the voltage from the conventional waveform—which is typically defined as a sinusoidal waveform with constant frequency and magnitude—are the main focus of the quality of voltage (QoV). The electric current's deviations from the usual waveform are the main focus of quality of current (QoC) [9]. However, relying solely on voltage or current to define PQ is imprecise because any divergence in voltage will result in a variation in current from its nominal value and vice versa [11]. Accordingly, PQ should combine both current and voltage qualities and is better described as technical limits that enable equipment to function in its prearranged way without significant operational losses to maintain its lifetime [9–12].

2. Power quality disturbances

PQ disturbances address various power system issues, including notching, voltage fluctuations, voltage flickers, harmonics, sub-harmonics, inter-harmonics, supra-harmonics, transients (oscillatory and impulsive), voltage sags, interruptions, voltage swells, imbalance, undervoltages, overvoltages, noise, and harmonics [9]. Even though each of them is a crucial topic in and of itself, the power system harmonics problem is regarded as the most significant and well-known PQ issue. This is because numerous studies have identified harmonics as the most serious cause of frequent PQ disturbances, impacting both consumers and utilities. Distribution system operators typically presume that harmonics and imbalances are the cause of PQ issues when they occur. **Figure 1** illustrates the various PQ issues [8–10]. Any PQ evaluation

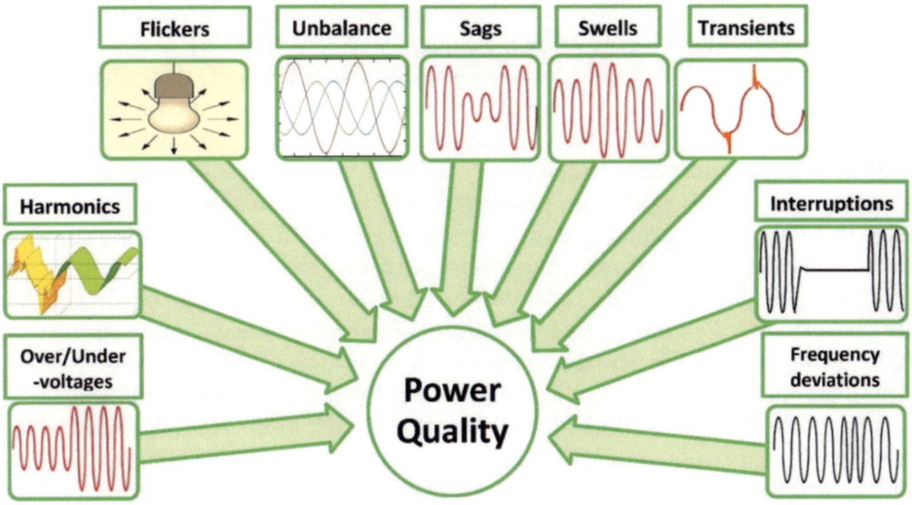

Figure 1.
Illustration of the common PQ issues.

procedure can be implemented with general steps, such as identifying the sources of the disturbances and considering potential harmonic management solutions up to the point of solution optimization. **Figure 2** examines a generic PQ diagnostic, assessment, and mitigation process [9, 13].

Depending on the region of the utility, standards and guidelines are typically used to categorize and identify problems. By enumerating the relevant characteristics, such as amplitude, frequency, spectrum, modulation, source impedance, notch depth, notch area, duration, rate of occurrence, and others, IEEE Std. 1159-2019 (as an example) classifies PQ or electromagnetic phenomenon.

The PQ indices were developed to provide a quantitative measure of the disruptiveness of disturbances; however, with the advancement of technology and changes in some systems' susceptibility to disturbances, the appropriateness of some PQ indices needs to be reevaluated. In addition, while some PQ indices have already been

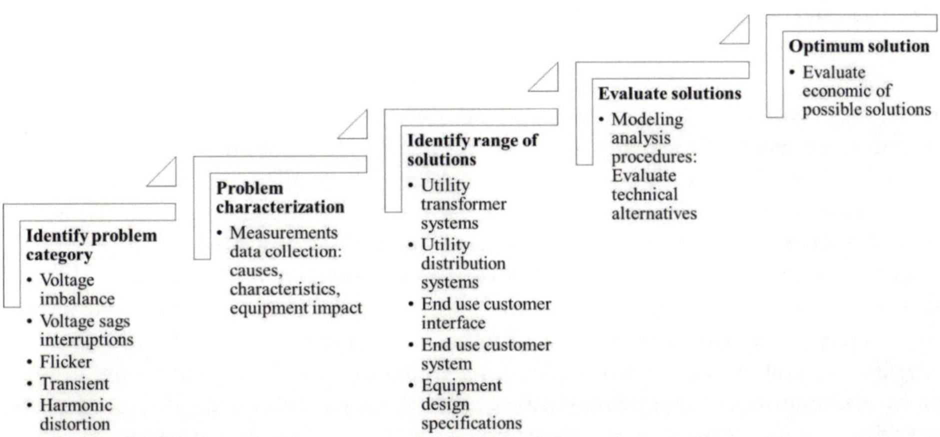

Figure 2.
PQ appraisal procedure.

defined or redefined in standards and their updates, others remain missing, especially for high and extra-high voltage or high-frequency systems [14, 15].

It is now crucial to develop criteria for limiting issues from PQ degradation due to the growing usage of nonlinear loads and renewable energy-based equipment with power electronic converters. The following are the primary PQ problems associated with DGs being connected to the power grid: DC injection, harmonics, voltage swells and sags, poor voltage regulation, power factor, flickering and fluctuating voltage, voltage imbalance, and prolonged interruptions. Further details regarding these problems, as well as technology and strategies for lessening the effects of DGs on PQ, are available in [16–18]. Egyptian Transmission/Distribution Codes, the Grid Connection Code for Solar Energy Plants, the International Electrotechnical Commission (IEC), the Institute of Electrical and Electronics Engineers (IEEE), the American National Standards Institute (ANSI), the National Institute of Standards and Technology (NIST), the National Fire Protection Association (NFPA), the European Committee for Electrotechnical Standardization (CENELEC), the National Electrical Manufacturers Association (NEMA), the Electric Power Research Institute (EPRI), Underwriters Laboratories (UL), and ESKOM for South African standards are just a few of the national and international organizations that have developed PQ standards. IEC and IEEE are the two prominent organizations that define PQ standards. PQ standards are referred to as electromagnetic-compatibility (EMC) standards by some. A number of EMC standards (series) and technical reports have been released by the IEC; the majority are included in the IEC 61000 series [19–24]. Numerous EMC standards that provide a thorough summary of IEEE PQ standards have been accepted by the IEC. The IEEE 519 recommendations are the most well-known substitute for the IEC standards in many countries [25, 26].

The goal of these standards is to restrict customers' access to harmonic distortion and the associated issues it causes, as well as the utility's voltage harmonic distortion boundaries. These guidelines divide the obligation of limiting harmonic propagation between utilities and end users. Customers and end users are typically in control of controlling the injection of harmonic currents, and utilities, regulators, and operators are in charge of figuring out the voltage distortion in the supply system at the point of common coupling (PCC).

3. Power system harmonics: definition

Typically, "any periodically distorted waveform can be represented as a sum of pure sine waves in which the frequency of each sinusoid is an integer multiple of the fundamental frequency of the distorted wave." The multiple-frequency has been named the fundamental's harmonic component, in which the so-called Fourier series refers to the summation of these sinusoids [27, 28]. If the fundamental frequency (f_1) is 50 Hertz, the 5th harmonic is 5*50 Hz or 250 Hz. Classically, amplitudes of the harmonic currents are expressed as a percentage of the fundamental current amplitude (I_f), so that $I_3 = I_f/3$, $I_5 = I_f/5$, $I_7 = I_f/7$, and so on. In line with the literature, in electric power system analysis, high-order harmonics above 25, i.e., the range from 25 to 50, are insignificant. It should be mentioned that harmonics above the 25th order are prevalent in telecommunication system studies. High-order harmonics might cause interference with power-electronic equipment; however, these harmonics are not critical to power system equipment [29]. It should be mentioned that recently supra-harmonics (distortion in the frequency range between 2 kHz and 150 kHz) [30] initiated to be considered in a few studies; however, they are not achievable in harmonic analysis in power systems due to the absence of legal restrictions for electric harmonic distortions in the very high-frequency range [31].

4. Harmonics sources and effects

One might refer to a load as nonlinear or non-ohm's law compliant when the current carried by the load is not proportionate to the applied voltage. This kind of load's current has a non-sinusoidal waveform or distorted current. This current distortion will generate a voltage distortion when there is a high impedance in the path between the source and the nonlinear load. On the contrary, one can say that loads with the current linearly proportional to the applied voltage are linear loads (linear relationship) [9, 10]. These days, the majority of loads are nonlinear in nature due to the widespread use of power-electronic-based components in power systems, even in our homes with our laptops, small appliances, fluorescent and light-emitting diode (LED) lamps, and printers. The load itself (design or component) and the nonlinear load's interaction with the distribution system determine how severe the harmonics produced by these loads are [25, 32]. As the primary sources of harmonic voltages and currents in power systems, several groups of power components can be grouped and organized as follows [9, 10]: transformers, electric motors, and generators (magnetic core-based equipment); induction furnaces, arc furnaces, and arc welders (equipment provides heating); and power-electronic-based devices. The way the power system is connected or composed is another classification. Furthermore, rather than the series-connected parts (linear series elements), the shunt-connected elements (loads, for example) are where the nonlinearities in the system arise. The magnetizing impedance (shunt-connected branch) of the well-known T model serves as the harmonic source inside a transformer, while the leakage impedance stands in for linear components. The most usual harmonic sources are—converters (inverters and rectifiers) within drives or renewables-based devices, slots and teeth field distribution in synchronous generator, power and distribution transformers' magnetizing circuits, rotating machines' excitation currents, printing machines, lamps (fluorescent, compact fluorescent, gas discharging lighting-low pressure/high pressure Sodium vapor, high-pressure mercury vapor and LED), flexible alternating current (AC) transmission systems, FACTS, and distributed FACTS (D-FACTS), uninterruptible power supplies, switch-mode power supplies, pulse modulation (or other forms) has been proposed for active power and voltage control in transmission circuits, electrolysis-based loads, converters usually used in variable speed drives (VSDs), converters used in grid-connected or islanded solar photovoltaic (PV) and wind systems, arc and induction welders, arc and conduction furnaces, and ovens used in electric heating, energy conservation device (soft starters, electronics ballast, and fan regulators), ballasts of the fluorescent discharge lamps, thyristor-controlled reactors, induction motors operating in or near to their saturation regions, converters in high-voltage direct current (HVDC) systems, UPSs, static VAR compensator and devices, and components in charging stations of electric vehicles (EVs). To sum up, most of these harmonic sources are power electronic-based devices.

The expected range of harmonics' impact is degradation in power system equipment's performance to their severe failure. The most common consequences of power system harmonics on the different electrical system sectors are explored in **Figure 3**.

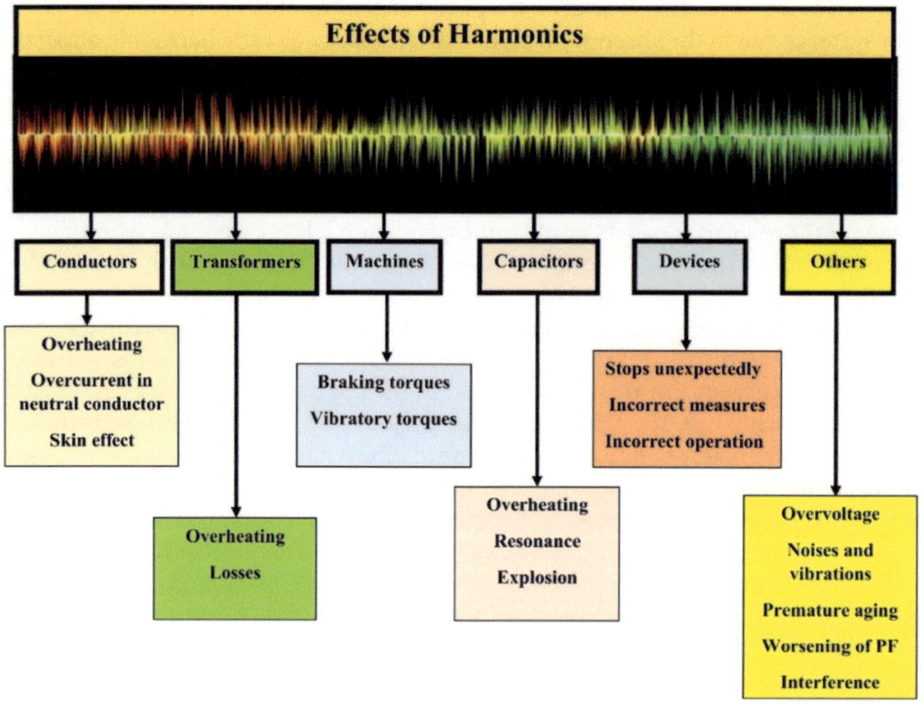

Figure 3.
Consequences of harmonics on components of the power system.

The most common harmonic problems in plants are summarized below [33, 34].

- Current flowing in neutral wires with an overheating problem
- High distorted currents will lead to excessive energy losses (thus high electricity consumption and costs)
- Unreasonable failure of equipment
- Motors' disturbance
- Overloading of frequency-dependent conductors
- Blow of fuses and mal-operation in the performance of protection devices
- Watt-hour metering's errors
- Interference with telecommunication systems (above the 25th order)
- Data loss in data-transmission networks
- Mal-operation in the performance of the control devices
- A voltage or current amplification (by series and parallel harmonic resonance)

- Harmonic instability (malfunctioning of voltage and malfunctioning of generator regulators)
- Noise in transformers
- Noise and vibrations in rotating machines
- Lockups of the programmable controllers

Usually, problems appear when a system's capacitance results in resonance with inductance at characteristic harmonic orders that intensely increases the distortion beyond the standard, acceptable values, as originate in industrial power systems because of the power factor correction capacitors that are frequently used and can cause a high degree of resonance severity or harmonic amplification. Such a privilege necessitates special considerations concerning harmonics filtering to avoid failures and nuisance tripping of fuses or breakers associated with capacitors [33, 34]. Adding filters that start with the lowest significant harmonic order (usually the third- or the fifth-order) is necessary to avoid harmonic resonance problems. If one wants to use a seventh harmonic order filter, one should introduce a fifth harmonic filter.

Further, analysis of the impedance-frequency dependencies for all reasonable operating contingencies should be done (in which a frequency scan should be conducted at each node if any harmonic source exists).

5. Harmonics mitigation

Since most electrical loads in use today are nonlinear, it is usually helpful to study reasonable harmonic solutions by systematically addressing electrical system-related concerns. When an issue arises, the fundamental solutions for harmonic control are to either add filters to sink the system's harmonic currents or stop them from entering the system, or reduce the harmonic currents generated by the load [9, 35].

Different attempts have been used to solve harmonics issues either to lessen their impacts on the power system or to reduce the harmonic distortion itself in the power grids, such as [36]:

- Derating transformers, motors, cables, and generators to be able to withstand the distorted over currents caused by harmonics
- The grounding of electrical equipment to cancel the severe 3rd harmonic and strengthen the neutral wire size
- Applying harmonic mitigation schemes such as active, passive, and hybrid filters
- Using multi-pulse converters

The harmonic filters can be classified as shunt filters or series filters based on the harmonic filter connected to the system. The shunt filters work by short-circuiting harmonic currents, which diverts the electric currents out of the arrangement. They must be placed as close as possible to the source of distortion. Shunt filtering is the most common way of filtering because of its economic aspects. Also, a shunt filter inclines to correct the load actual and displacement power factors and mitigate harmonic currents [29]. The other way is to put on a series-connected filter that helps in blocking harmonic currents. Nevertheless, series-connected filters must be planned to withstand the rated line current as they are connected in series with the system. Also, series-connected filters may produce substantial power losses because of the high currents. Given the high cost of the series-connected filters, the most real-world applied method is shunt-connected filters. Several contributions are dedicated to determining the most suitable location of the filters in electrical power networks. Filters at appropriate places (close to the source of harmonic generation) can be applied to mitigate considerable harmonic currents at the start, and the harmonics propagation to the common coupling point (PCC) is considerably reduced.

On the contrary, the harmonic flow occurs when the filters are far from the harmonic-producing loads. Harmonic filters are also categorized into three broad categories: passive, active, and hybrid active/passive filters. **Figure 4** presents the primary ways of connecting harmonic filters at the PCC.

Passive filters comprise inductive (L), capacitive (C), and resistive (R) components arranged and lumped together in precise configurations to regulate harmonics. They are commonly used in practice because they are considerably inexpensive compared with other active/hybrid filters. Nonetheless, they have the disadvantages of negatively interrelating with power systems and threatening the utility (source) and the loads (within the plant or neighbors) by harmonic resonance hazards [29]. Also, their filtering performance is sensitive to the variation of the source impedance [37].

The active harmonic filtering method was a reasonably innovative methodology for eliminating harmonics compared to the passive filtering techniques. Compact constructed active filters provide reliable system performance with good harmonic lessening. However, they are based on power electronic components; thus, they are more costly than passive filters. The basic concept of active filters introduces equal magnitudes of the current/voltage harmonics generated by nonlinear loads with 180° phase angle difference; consequently, they cancel each other when their phasor is summed. In addition, they do not resonate with the system [9, 34, 35, 38, 39]. By definition, active filters are designed based on converter type, topology, and the number of phases. The converter type can be either a current source-based inverter that employs an inductor to store energy or a voltage source-based inverter that employs a capacitor to store energy [9, 34, 35, 38, 39]. The arrangement can be shunt-connected types, series-connected types, or a combination of both connections. Active filters have frequency limitations, cannot withstand large currents and are sensitive to noise. Also, they have problems with high-power ratings (>0.5 MVA).

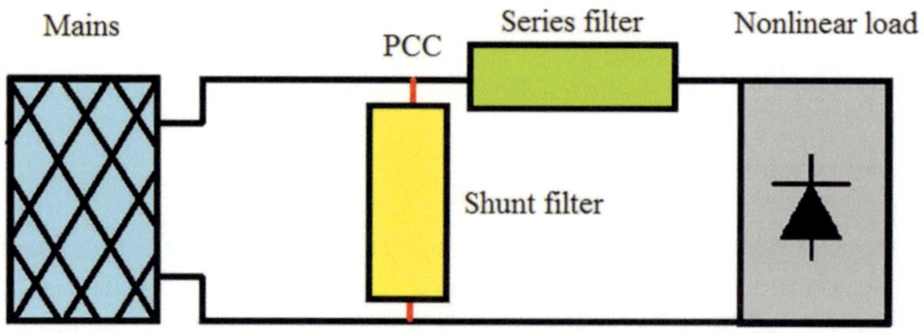

Figure 4.
Basic connections of harmonic filters.

Nowadays, both active and passive filters can be used in the presence of multiple pulse converters governed by *harmonics* = *Integer* * *pulse* ± 1, where *pulse* is the pulse number and *Integer* = 1, 2, 3, etc. Harmonic current distortions of 6, 12, and 18 pulse converters are higher (in THD percentage) than 80%, 15%, and 12%, respectively. Multiple pulse converters of THD less than 5% are expensive to the manufacturers.

A straightforward technique to decrease harmonics is to increase the pulse numbers of converters. The lessening of harmonics with the increase of pulse number is guaranteed. Disadvantages of multiple pulse converters include sensitivity to voltage imbalance, optimal cancelation only with symmetric drive loading (they do not operate well with even harmonics), and not being easy to retrofit [23, 40].

Hybrid filters benefit from passive and active filters through series or parallel combinations. A passive filter helps to reduce the rating of the used active filter and its function in harmonic mitigation and improvement of power factors. The role of the active filter is to isolate the generated harmonics of both load and utility. Other harmonic management solutions concerning harmonic correction equipment types, such as the neutral blocking filters or the zigzag transformers, are solutions to eliminate the 3rd harmonic current from the load. Typically, they are suitable for computer/switch-mode power supplies. Other solutions have a kind of immunity to harmonic distortion, such as the oversized neutral/derated transformers, *K*–rated transformers, and phase shifting. They are more suitable for fresh/new designs in the planning stage; they do not have power factor correction benefits [41, 42]. They are much more superlative for commercial applications than industrial applications [42].

6. Renewable energy resources and their PQ issues

Within the framework of sustainable development, adding renewable energy sources to distribution or transmission systems has several advantages. These include promoting the use of green energy, diversifying energy sources, reducing greenhouse gas emissions, gaining political advantages, fostering social development, and providing economic support. There are also numerous technical advantages, such as improved power quality, reduced power loss, improved voltage, and increased load stability.

Distributed generators can be a renewable or non-renewable source of generation and can be networked (grid-connected) or act as a stand-alone system. Due to their low investment costs and small sizes, DGs show an imperative role in modern energy system planning [43]. DGs can be classified based on several issues such as [44, 45]—generated power (AC or DC); technology (Renewable (non-fossil fuel-based and non-renewable (fossil fuel-based)); supply duration (long duration, short duration, moderate but unsteady duration); capacity (micro decentralized DGs (1 W–5 kW), small decentralized or centralized DGs (5 kW–5 MW), medium centralized DGs (5 MW–50 MW), and large centralized DGs (50 MW–300 MW); grid interface (Inverter-based and non-inverter-based DGs)); power flow control (set to constant power factor for small DGs, and the bus at which the DG is connected is treated as a PQ bus, or set to constant voltage for large DGs and the bus at which the DG is connected is treated as a PV bus); and power delivering capability (deliver only active power at unity PF, deliver only reactive power at zero PF, DGs deliver active power but consumes reactive power, or deliver both active and reactive powers).

The energy flow and voltage conditions at customers and utility equipment are greatly impacted by the addition of DGs to distribution networks. Depending on the distribution networks, DG characteristics, and operating parameters, these effects could be either beneficial or detrimental [46, 47]. However, if renewable energy sources are not allocated properly, their unchecked growth could cause issues for power systems. Some common issues include overloading transformers, increasing power loss, malfunctioning or failing protection schemes, excessive harmonic distortion levels brought on by the combination of nonlinear loads and inverter-based renewable energy sources, as well as over- and under-voltages. When the rating of DGs surpasses the maximum permitted degree of penetration of renewables, these challenges arise.

7. Hosting capacity (HC): problems and solutions

When the system surpasses the maximum allowable hosting (HC) capacity criteria, DGs integration issues arise [47]. To decide on the addition of renewables, various customary rules of thumb were employed in the past. Instantaneous penetration (IP) was the term used to describe a definition that was previously introduced and was similar to the HC. IP has been described as the ratio of the output of renewable energy to the power of the system load within a given time or brief interval. It was not, however, frequently applied as the HC definition. These days, the inclusion of renewables can be determined by HC. Because of its significance, HC has been integrated into popular simulation programs as Siemens, CYME, ETAP, DIGSilent, and EPRI (DRIVE). **Figure 5** examines a case study of determining HC in relation to penetration level using a generic performance metric, along with related issues and solutions. The type of problem that has emerged determines these performance measures. As can be seen from the figure, overvoltage, overloading and subsequent power loss complexities, PQ concerns, and protection issues are the four primary problematic issues [48, 49].

To guarantee that the power system functions satisfactorily, the HC approach compiles the technical limitations implemented by operators and customers.

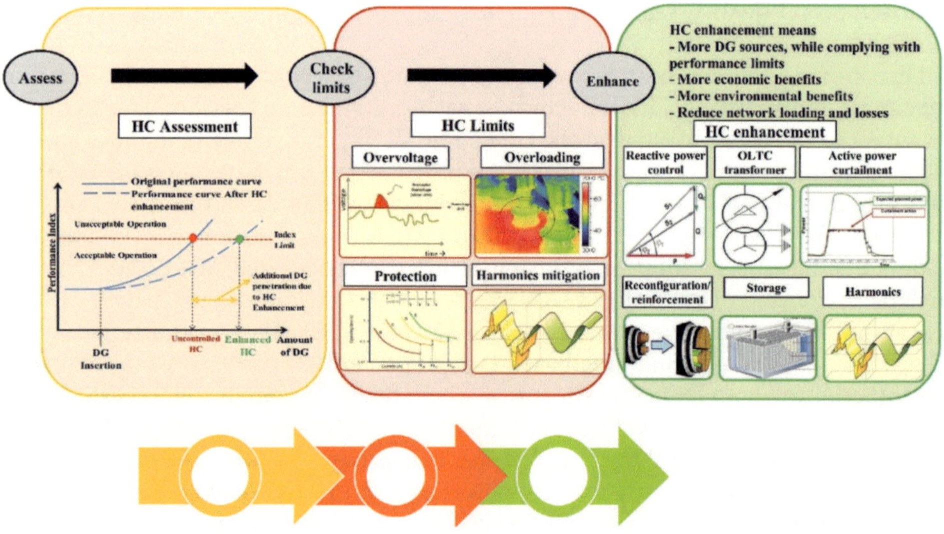

Figure 5.
HC determination by utilizing common solutions, troublesome problems, and a general performance metric.

This indicates more to the HC calculation than a single, static calculation based on a single performance parameter. On the other hand, HC would be determined for a number of performance indicators, including PQ, thermal overload capacity, voltage variations and frequency fluctuations, system stability, and others.

Ismael et al. provide a thorough overview of HC's advancements, assessment procedures, and improved technology in [47]. Decisions from real case studies, power quality markets, and the practical experience (rule-of-thumb) of distribution system operators are given and discussed [47]. The authors in [50] also presented a detailed analysis of HC—theory and its influences on power networks, challenges, and solutions.

In the literature, numerous strategies have been put into practice to raise the HC of distribution systems. The most popular methods are:

- Renewables curtailment (when investors and operators are asked to reduce the amount of renewable energy they produce in order to maintain system working limits) [51]. To achieve the best possible power curtailment, system and renewable plant operators must have creative communication techniques and facilities.

- Use of energy storage devices to boost the system performance and permit its consistent act without renewables ceiling [52, 53]. Comparable benefits are provided by energy storage, which are challenging to provide with other approaches. But the primary drawback is the high cost of energy storage.

- Reconfiguring of nodes of RDS changes the status of the operating switches by controlling tie-lines, sectionalizes, and soft open points can reduce power losses, transfer loads between feeders, improve the nodes' voltage profile, and improve HC, PQ, and reliability [54] of the system. Reconfiguration can be employed in the planning stage of power systems (called static reconfiguration) or in the operation phase of power systems (called dynamic reconfiguration). The static reconfiguration can considerably enhance HC, but the dynamic reconfiguration can only improve HC in case of the availability of an adequate number of controlled switches.

- Use of harmonic mitigation techniques, as proposed in this thesis, such as shunt capacitor banks [55], static VAR compensators [56], D-FACTs, and harmonic filters (dominantly passive or hybrid filters) to lessen harmonic distortion, support reactive power, correct the power factor and improve PQ performance of power systems operating under non-sinusoidal conditions [48, 57].

- Use of voltage regulators/conditioners, reactive power compensators, and OLTCs (on-load tap changers) to improve voltage profiles and, sequentially, support reactive power and enhance the HC of the system [58].

- Reinforcement of weakened or congested systems [59] can also be made to increase HC. In this regard, reinforcement means using machines with a higher rating, larger conductor sizes that have lower electrical resistance) or using efficient equipment. This is considered one of the practical techniques used in congested systems: support the voltage profile, achieve better hosting capacity, relieve the electrical system congestions, and reduce network losses.

- Interbreeding of diverse solutions to attain the best possible HC values in intelligent power grids [60], especially in severely deteriorated systems or in projects with high budgets.

To summarize, the following steps have to be performed to enhance the HC of a power system:

- Evaluate the initial HC within the network.
- Check the operational limits according to the international standards or national practice codes.
- Employ a proper HC enhancement technique.
- Re-evaluate the new HC value.

The complete HC procedure—evaluation and improvement—is summarized in **Figure 6**.

Analytic HC calculation procedure means a precise process is done by iteratively increasing the penetration of renewables in a well-defined step at a carefully chosen bus, carrying out load flow calculations, and inspecting the operating limits at each iteration until they exceed the acceptable values; henceforth, finding the HC of that bus. Then, another bus is selected in sequence, and the same process is repeated till all the system busses are examined. This may suffer from the computational burden.

The stochastic HC calculation procedure means developing multiple scenarios in a probabilistic manner to overcome the uncertainties of the problem. It should be mentioned that this method is time-dependent based on the accuracy levels considered. Finally, a streamlined HC calculation procedure provides easy, quick screens that assist the operator in deciding whether it is required to make further detailed studies or not. However, this method suffers from accuracy issues, particularly in complex systems.

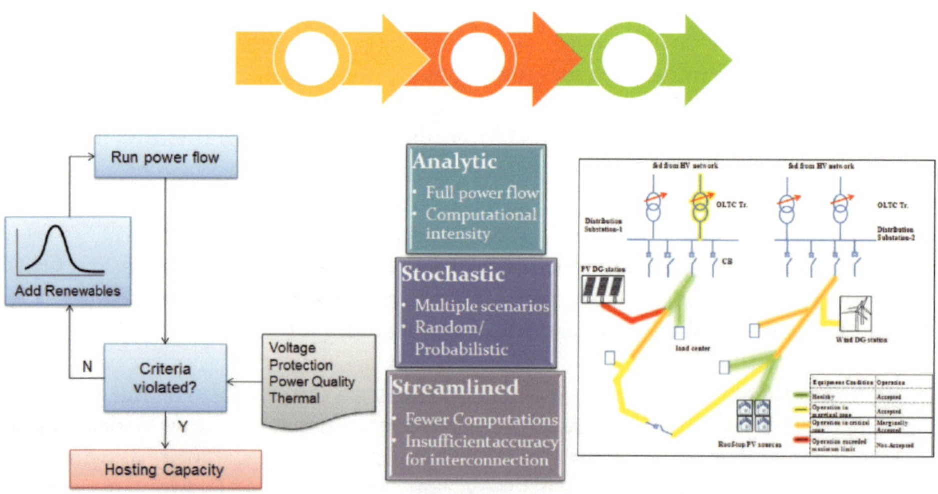

Figure 6.
HC calculation procedure.

In the past, harmonic distortion was not one of the main interests of operators because of the assumption, at that time, that distributed generation units are harmonic-free. Later, it was realized that this assumption is not true as the combination of harmonics between distributed generation units and non-linear loads could create severe problems.

The authors in [61] used a passive harmonic filter, a *C*-type passive filter, to maximize the HC of a network that utilizes dispersed generators (PV units) in a harmonically distorted distribution system using the genetic algorithm. From the analysis presented, it was apparent that the system's HC goes down with the rise in the grid-side voltage-distortion and the nonlinear load level. The HC level was affected more by the non-linear load level than grid-side voltage distortion. Also, the same authors used a single-tuned filter in [49] to do the same, and similar conclusions were figured out in [49] to validate that harmonic filters can solve the problems arising from harmonic distortion and enhance HC at the same time. Ghaffarzadeh and Sadeghi in [62] presented an effective method for the simultaneous settlement of inverter-based DG systems and capacitors because of harmonic distortion. However, HC was not discussed in detail in that work. Further, the authors in [63] investigated the use of passive filters in different single-objective optimization problems in the context of HC improvement along with THD lessening. The analysis revealed that HC improvement and voltage THD lessening were conflicting in optimization and that a multi-objective optimization may be needed to solve the problem. Similarly, a few other works are also reported, making an effort to improve HC using passive harmonic filtering techniques [48, 64], but in a single-objective optimization framework. Further, hybrid harmonic distortion mitigation is presented in [65] to improve distributed generation-based systems' HC in harmonic-polluted conditions, in which the hybrid filter was a better substitute to realize a higher penetration level of renewables than purely passive filters, regardless of the cost of filtering and the limited rating of the active filters used. To summarize, the main focus of researchers is clearly moving to empower distribution systems with highly penetrated renewables while offering multi-functionality facilities (a trade-off between different goals). Accordingly, multi-objective optimization should be used to improve the HC while limiting harmonics and improving the PQ performance of such distorted systems.

8. Summary

- The integration of distributed generation and renewable energy sources into distribution systems has sparked a growing interest in understanding the HC of these systems while ensuring reliability, PQ, and sustainability.

- Traditional assumptions that distributed generation units are harmonic-free have been challenged as it became evident that the combination of harmonics between these units and nonlinear loads can lead to severe PQ problems. To overcome the uncertainties associated with HC calculations, researchers have developed stochastic procedures that consider multiple probabilistic scenarios. These approaches provide a more comprehensive understanding of system performance and aid in decision-making processes.

- Harmonic distortion, once not a primary concern for operators, has now emerged as a critical issue. Passive harmonic filters have been employed to maximize HC and mitigate harmonic distortion. The analysis presented in various studies has demonstrated that the HC of a distribution system decreases with increasing grid-side voltage distortion and nonlinear load levels. Moreover, it has been observed that nonlinear load levels have a more significant impact on HC than grid-side voltage distortion. These findings highlight the effectiveness of harmonic filters in addressing the problems arising from harmonic distortion and enhancing HC simultaneously. However, optimizing HC and improving power quality pose complex challenges. Single-objective optimization approaches may not suffice, as there can be conflicting objectives. Researchers have found that HC improvement and voltage total harmonic distortion (THD) reduction tend to be conflicting goals in optimization. Consequently, multi-objective optimization techniques have been proposed to strike a balance between HC improvement, harmonic reduction, and voltage THD lessening.

- The drive toward sustainability has urged researchers to seek solutions that enable distribution systems to accommodate highly penetrated renewables while offering multi-functionality. Hybrid harmonic distortion mitigation techniques have emerged as a promising approach. By combining passive and active filters, these techniques can achieve higher levels of renewable energy penetration, regardless of the cost of filtering and limitations of the active filters used. This integration of renewables and hybrid filtering not only enhances HC but also contributes to the long-term sustainability goals of the system.

In summary, the convergence of HC, reliability, PQ, and sustainability requires the adoption of advanced techniques, including stochastic HC calculation, passive and hybrid harmonic filtering, and multi-objective optimization. By considering these factors together, distribution systems can effectively accommodate higher levels of renewable generation, ensure high-quality power supply, improve system reliability, and contribute to long-term sustainability objectives. These research efforts pave the way for the empowerment of distribution systems with renewable energy while maintaining the necessary functionality and performance standards.

Author details

Muhyaddin Rawa[1,2], Ziad M. Ali[3,4] and Shady H.E. Abdel Aleem[5*]

1 Smart Grids Research Group, Center of Research Excellence in Renewable Energy and Power Systems, King Abdulaziz University, Jeddah, Saudi Arabia

2 Faculty of Engineering, Department of Electrical and Computer Engineering, King Abdulaziz University, Jeddah, Saudi Arabia

3 Electrical Engineering Department, College of Engineering, Prince Sattam bin Abdulaziz University, Wadi Addawaser, Saudi Arabia

4 Aswan Faculty of Engineering, Electrical Engineering Department, Aswan University, Aswan, Egypt

5 Department of Electrical Engineering, Institute of Aviation Engineering and Technology, Giza, Egypt

*Address all correspondence to: shady.abdelaleem@iaet.edu.eg

References

[1] Omar AI, Abdel Aleem SHE, El-Zahab EEA, Algablawy M, Ali ZM. An improved approach for robust control of dynamic voltage restorer and power quality enhancement using grasshopper optimization algorithm. ISA Transactions. 2019;**95**:110-129. DOI: 10.1016/j.isatra.2019.05.001

[2] Gumilar L, Cahyani DE, Afandi AN, Monika D, Rumokoy SN. Optimalization harmonic shunt passive filter using detuned reactor and capacitor bank to improvement power quality in hybrid power plant. In: AIP Conference Proceedings. Vol. 2217. AIP Publishing: The 5th International Conference on Industrial, Mechanical Electrical, and Chemical Engineering 2019 (ICIMECE 2019); 2020. p. 030003. DOI: 10.1063/5.0000710

[3] Sakar S, Balci ME, Aleem SHEA, Zobaa AF. Hosting capacity assessment and improvement for photovoltaic-based distributed generation in distorted distribution networks. In: 2016 IEEE 16th International Conference on Environment and Electrical Engineering (EEEIC). IEEE; 2016. pp. 1-6. DOI: 10.1109/EEEIC.2016.7555515

[4] Wang X, Blaabjerg F. Harmonic stability in power electronic-based power systems: Concept, modeling, and analysis. IEEE Transactions on Smart Grid. 2019;**10**:2858-2870. DOI: 10.1109/TSG.2018.2812712

[5] Zobaa AF, Abdel Aleem SHE, Ismael SM, Ribeiro PF. Hosting Capacity for Smart Power Grids. Cham, Switzerland: Springer; 2020. pp. 1-254. DOI: 10.1007/978-3-030-40029-3

[6] Institute of Electrical and Electronics Engineers. IEEE Std 1159—IEEE Recommended Practice for Monitoring Electric Power Quality. Vol. 2009. New York City, United States: IEEE; 2009. DOI: 10.1109/IEEESTD.2009.5154067

[7] Heydt GT. Electric power quality. In: The Electrical Engineering Handbook. Amsterdam, Netherlands: Elsevier, Academic Press; 2005. pp. 805-810. DOI: 10.1016/B978-012170960-0/50059-1

[8] Dugan RC, McGranaghan MF, Santosa S, Beaty HW. Electric Power Systems Quality: Distributed Generation and Power Quality. 2nd ed. New York: McGraw-Hill; 2002

[9] Bollen MHJ. What is power quality? Electric Power Systems Research. 2003;**66**:5-14. DOI: 10.1016/S0378-7796(03)00067-1

[10] Fuchs E, Masoum M. Power Quality in Power Systems and Electrical Machines. Amsterdam, Netherlands: Elsevier, Academic Press; 2008. DOI: 10.1016/B978-0-12-369536-9.X5001-3

[11] Baretich MF. Electrical power. In: Clinical Engineering Handbook. 2nd ed. Amsterdam, Netherlands: Elsevier, Academic Press; 2019. pp. 667-669. DOI: 10.1016/B978-0-12-813467-2.00093-6

[12] Abdel Aleem SHE, Zobaa AF, Abdel Aziz MM. Optimal C-type passive filter based on minimization of the voltage harmonic distortion for nonlinear loads. IEEE Transactions on Industrial Electronics. 2012;**59**:281-289. DOI: 10.1109/TIE.2011.2141099

[13] Bollen M, Beyer Y, Styvactakis E, Trhulj J, Vailati R, Friedl W. A European benchmarking of voltage

quality regulation. In: International Conference on Harmonics and Quality of Power. New York City, United States: IEEE; 2012. pp. 45-52. DOI: 10.1109/ICHQP.2012.6381171

[14] Kandil MS, Farghal SA, Elmitwally A. Refined power quality indices. IEE Proceedings – Generation, Transmission and Distribution. 2001;**148**:590-596. DOI: 10.1049/ip-gtd:20010515

[15] Heydt GT, Jewell WT. Pitfalls of electric power quality indices. IEEE Transactions on Power Delivery. 1998;**13**:570-576. DOI: 10.1109/61.660930

[16] Patil KR, Karnik SR, Raju AB. Impacts of distributed generation on power system stability. In: 2021 International Conference on Intelligent Technologies (CONIT). New York City, United States: IEEE; 2021. pp. 1-6. DOI: 10.1109/CONIT51480.2021.9498452

[17] Van TV, Driesen J. Distributed generation and power quality. In: Handbook of Power Quality. Vol. 145. California, USA: John Wiley & Sons, Ltd.; 2008. pp. 521-528. DOI: 10.1002/9780470754245.ch16

[18] Mittal D. Classification of power quality disturbances in electric power system: A review. IOSR Journal of Electrical and Electronics Engineering. 2012;**3**:06-14. DOI: 10.9790/1676-0350614

[19] Halpin SM. Comparison of IEEE and IEC harmonic standards. In: IEEE Power Engineering Society General Meeting, 2005. Vol. 3. San Francisco, California, USA: IEEE; 2005. pp. 2214-2216. DOI: 10.1109/pes.2005.1489688

[20] Prudenzi A, Grasselli U, Lamedica R. IEC Std. 61000-3-2 harmonic current emission limits in practical systems: Need of considering loading level and attenuation effects. In: Proceedings of IEEE/PES Transmission and Distribution Conference. Vol. 1. Vancouver, BC, Canada: Power Engineering Society Summer Meeting; Conference Proceedings (Cat. No.01CH37262) 2001. pp. 277-282. DOI: 10.1109/pess.2001.970026

[21] International Electrotechnical Commission. Electromagnetic Compatibility (EMC) Part 3-6: Limits — Assessment of Emission Limits for the Connection of Distorting Installations to MV, HV and EHV Power Systems. Geneva, Switzerland: International Electrotechnical Commission (IEC). 2008. Iec 61000-3-6 2008

[22] McGranaghan M, Beaulieu G. Update on IEC 61000-3-6: Harmonic emission limits for customers connected to MV, HV, and EHV. In: Proceedings of 2005/2006 IEEE/PES Transmission and Distribution Conference and Exhibition, Dallas, TX, USA. 2006. pp. 1158-1161. DOI: 10.1109/TDC.2006.1668668

[23] Kalair A, Abas N, Kalair AR, Saleem Z, Khan N. Review of harmonic analysis, modeling and mitigation techniques. Renewable and Sustainable Energy Reviews. 2017;**78**:1152-1187. DOI: 10.1016/j.rser.2017.04.121

[24] IEC61000-3-2. Electromagnetic Compatibility (EMC) - Part 3-2: Limits for Harmonic Current Emissions. Geneva, Switzerland: International Electrotechnical Commission (IEC); 61000-3-2; 2005

[25] Aleem SHEA, Elmathana MT, Zobaa AF. Different design approaches of shunt passive harmonic filters based on IEEE Std. 519-1992 and IEEE Std. 18-2002. Recent Patents on Electrical and Electronics Engineering. 2013;**6**:68-75. DOI: 10.2174/2213111611306010009

[26] IEEE. 519-2014. IEEE Recommended Practice and Requirements for Harmonic

Control in Electric Power Systems | IEEE Standard | IEEE Xplore. Vol. 2014. New York City, United States: IEEE; 2017

[27] Bollen MHJ, Gu IYH. Signal Processing of Power Quality Disturbances. Canada: John Wiley & Sons, Inc.; 2005. DOI: 10.1002/0471931314

[28] Ellis R, Eng P. Power System Harmonics–a Reference Guide to Causes, Effects and Corrective Measures: An Allen-Brandley Ser Issues Answers. Canada: Rockwell International Corporation; 2001. pp. 1-14

[29] Singh A. Harmonic solutions to commercial and industrial electrical power systems. SAMRIDDHI A Journal of Physical Sciences Engineering and Technology. 2015;**6**:33-42. DOI: 10.18090/samriddhi.v6i1.1551

[30] Bollen MHJ, Rönnberg SK. Hosting capacity of the power grid for renewable electricity production and new large consumption equipment. Energies. 2017;**10**:1325. DOI: 10.3390/en10091325

[31] Aleem SHEA, Zobaa AF, Balci ME, Ismael SM. Harmonic overloading minimization of frequency-dependent components in harmonics polluted distribution systems using Harris hawks optimization algorithm. IEEE Access. 2019;**9**:100824-100837. DOI: 10.1109/ACCESS.2019.2930831

[32] Schneider Electric. Electrical Installation Guide According to IEC International Standard. Conference Record of 2007 Annual Pulp and Paper Industry Technical Conference (The Williamsburg Lodge, Williamsburg, Virginia). France: Schneider Electric, Les Deux-Ponts; 2009. p. 467

[33] Blooming TM, Carnovale DJ. Capacitor application issues. In: Conference Record of 2007 Annual Pulp and Paper Industry Technical Conference, The Williamsburg Lodge, Williamsburg, Virginia. 2007. pp. 178-190. DOI: 10.1109/PAPCON.2007.4286298

[34] Ogheneovo JD. Issues of power quality in electrical systems. International Journal of Energy and Power Engineering. 2016;**5**:148. DOI: 10.11648/j.ijepe.20160504.12

[35] Baggini A. Handbook of Power Quality. California, USA: John Wiley & Sons, Inc.; 2008. DOI: 10.1002/9780470754245

[36] Ghorbani MJ, Mokhtari H. Impact of harmonics on power quality and losses in power distribution systems. International Journal of Electrical and Computer Engineering. 2015;**5**:166-174. DOI: 10.11591/ijece.v5i1.pp166-174

[37] Abdel Aleem SHE, Ibrahim AM, Zobaa AF. Harmonic assessment-based adjusted current total harmonic distortion. Journal of Engineering. 2016;**2016**:64-72. DOI: 10.1049/joe.2016.0002

[38] Dugan RC, McGranaghan MF, Wayne BH. Electrical power systems quality. Choice Reviews Online. 1996;**34**:34-0322-34-0322. DOI: 10.5860/choice.34-0322

[39] Singh B, AlHaddad K, Chandra A. A review of active filters for power quality improvement. IEEE Transactions on Industrial Electronics. 1999;**46**:960971. DOI: 10.1109/41.793345

[40] Kazem HA. Harmonic mitigation techniques applied to power distribution networks. Advances in Power Electronics. 2013;**2013**:1-10. DOI: 10.1155/2013/591680

[41] Carnovale DJ. Applying Harmonic Solutions to Commercial and Industrial

Power Systems. Moon Township, PA: Eaton, Cutler-Hammer; 2003

[42] Carnovale DJ. Correct Application of Power Factor Correction Equipment. KVAR Engineering Services Limited Appliances, Electrical, and Electronics Manufacturing (Darlington, Co Durham, UK). n.d. Available from: http://www.kvarengineering.co.uk/correct-application-of-power-factor-correction-equipment/

[43] Bansal R. Handbook of Distributed Generation: Electric Power Technologies, Economics and Environmental Impacts. Cham: Springer International Publishing; 2017. DOI: 10.1007/978-3-319-51343-0

[44] Masoum MAS, Fuchs EF. Power Quality in Power Systems and Electrical Machines: Second Edition. Amsterdam, Netherlands: Elsevier; 2015. DOI: 10.1016/C2013-0-18758-2

[45] Diaaeldin I, Aleem SA, El-Rafei A, Abdelaziz A, Zobaa AF. Optimal network reconfiguration in active distribution networks with soft open points and distributed generation. Energies. 2019;**12**:4172. DOI: 10.3390/en12214172

[46] Diaaeldin IM, Abdel Aleem SHE, El-Rafei A, Abdelaziz AY, Zobaa AF. Hosting capacity maximization based on optimal reconfiguration of distribution networks with optimized soft open point operation. In: Smart Power Grids. Cham, Switzerland: Springer; 2020. pp. 179-193. DOI: 10.1007/978-3-030-40029-3_8

[47] Ismael SM, Abdel Aleem SHE, Abdelaziz AY, Zobaa AF. State-of-the-art of hosting capacity in modern power systems with distributed generation. In: Renewable Energy. Vol. 130. Amsterdam, Netherlands: Elsevier; 2019. pp. 1002-1020. DOI: 10.1016/j.renene.2018.07.008

[48] Bajaj M, Kumar SA. Hosting capacity enhancement of renewable-based distributed generation in harmonically polluted distribution systems using passive harmonic filtering. Sustainable Energy Technologies and Assessments. 2021;**44**:101030. DOI: 10.1016/j.seta.2021.101030

[49] Sakar S, Balci ME, Abdel Aleem SHE, Zobaa AF. Integration of large- scale PV plants in non-sinusoidal environments: Considerations on hosting capacity and harmonic distortion limits. Renewable and Sustainable Energy Reviews. 2018;**82**:176-186. DOI: 10.1016/j.rser.2017.09.028

[50] Azibek B, Abukhan A, Nunna HSVSK, Mukatov B, Kamalasadan S, Doolla S. Hosting capacity enhancement in low voltage distribution networks: Challenges and solutions. In: IEEE International Conference on Power Electronics, Smart Grid and Renewable Energy (PESGRE2020). New York City, United States: IEEE; 2020. pp. 1-6. DOI: 10.1109/PESGRE45664.2020.9070466

[51] Etherden N, Bollen MHJ. Increasing the hosting capacity of distribution networks by curtailment of renewable energy resources. In: IEEE PES Trondheim PowerTech Power Technol. A Sustain. Soc. POWERTECH 2011. New York City, United States: IEEE; 2011. pp. 1-7. DOI: 10.1109/PTC.2011.6019292

[52] Jayasekara N, Masoum MAS, Wolfs PJ. Optimal operation of distributed energy storage systems to improve distribution network load and generation hosting capability. IEEE Transactions on Sustainable Energy. 2016;7:250-261. DOI: 10.1109/TSTE.2015.2487360

[53] Rawa M, Abusorrah A, Al-Turki Y, Mekhilef S, Mostafa MH, Ali ZM, et al.

Optimal allocation and economic analysis of battery energy storage systems: Self-consumption rate and hosting capacity enhancement for microgrids with high renewable penetration. Sustainability. 2020;**12**:1-25. DOI: 10.3390/su122310144

[54] Ali ZM, Diaaeldin IM, Abdel Aleem SHE, El-Rafei A, Abdelaziz AY, Jurado F. Scenario-based network reconfiguration and renewable energy resources integration in large-scale distribution systems considering parameters uncertainty. Mathematics. 2021;**9**:1-31. DOI: 10.3390/math9010026

[55] Xiao J, Li Y, Tan Y, Qiao X, Cao Y, Zhang Y, et al. MILP model for hosting capacity assessment of distributed generation in distribution networks considering ZIP load model. In: iSPEC 2019—2019 IEEE Sustainable Power Energy Conference: Grid Modernization of Energy Revolution, Proceeding. New York City, United States: IEEE; 2019. pp. 1551-1555. DOI: 10.1109/iSPEC48194.2019.8975043

[56] Xu X, Li J, Xu Z, Zhao J, Lai CS. Enhancing photovoltaic hosting capacity—A stochastic approach to optimal planning of static var compensator devices in distribution networks. Applied Energy. 2019;**238**:952-962. DOI: 10.1016/j.apenergy.2019.01.135

[57] Ismael SM, Aleem SHEA, Abdelaziz AY, Zobaa AF. Probabilistic hosting capacity enhancement in non-sinusoidal power distribution systems using a hybrid PSOGSA optimization algorithm. Energies. 2019;**12**:185-217. DOI: 10.3390/en12061018

[58] Wang S, Chen S, Ge L, Wu L. Distributed generation hosting capacity evaluation for distribution systems considering the robust optimal operation of OLTC and SVC. IEEE Transactions on Sustainable Energy. 2016;**7**:1111-1123. DOI: 10.1109/TSTE.2016.2529627

[59] Ismael SM, Abdel Aleem SHE, Abdelaziz AY, Zobaa AF. Practical considerations for optimal conductor reinforcement and hosting capacity enhancement in radial distribution systems. IEEE Access. 2018;**6**:27268-27277. DOI: 10.1109/ACCESS.2018.2835165

[60] Zobaa AF, Abdel Aleem SHE, Ismael SM, Ribeiro PF. Hosting Capacity for Smart Power Grids. Cham: Springer International Publishing; 2020. DOI: 10.1007/978-3-030-40029-3

[61] Sakar S, Balci ME, Abdel Aleem SHE, Zobaa AF. Increasing PV hosting capacity in distorted distribution systems using passive harmonic filtering. Electric Power Systems Research. 2017;**148**:74-86. DOI: 10.1016/j.epsr.2017.03.020

[62] Ghaffarzadeh N, Sadeghi H. A new efficient BBO based method for simultaneous placement of inverter-based DG units and capacitors considering harmonic limits. International Journal of Electrical Power & Energy Systems. 2016;**80**:37-45. DOI: 10.1016/j.ijepes.2016.01.030

[63] Mohsen, Ismael S, Abdel Aleem SHE, Abdelaziz AY. Hosting capacity enhancement of electrical distribution systems under sinusoidal and non-sinusoidal conditions. In: Twentieth International Middle East Power Systems Conference MEPCON 2018—Proceedings. New York City, United States: IEEE; 2019. pp. 168-173. DOI: 10.1109/MEPCON.2018.8635218

[64] Ismael SM, Aleem SHEA, Abdelaziz AY, Zobaa AF. Probabilistic hosting capacity enhancement in non-sinusoidal power distribution systems using a hybrid PSOGSA optimization

algorithm. Energies. 2019;**12**:1018. DOI: 10.3390/en12061018

[65] Bajaj M, Singh AK. Performance assessment of hybrid active filtering technique to enhance the hosting capacity of distorted grids for renewable energy systems. International Journal of Energy Research. 2022;**46**:2783-2809. DOI: 10.1002/er.7345

Chapter 2

Impact of Harmonics on the Electrical Network Distribution

Abdourahimoun Daouda

Abstract

Today, electronic devices are used in all areas of human activities. However, these equipments generate a signal distortion of the electrical parameters of the voltage source that supplies these non-linear loads. When the level of distortion reaches a certain threshold value, it causes a premature deterioration of the electrical loads and even of the power supply devices (transformers, circuit breakers, etc.). So, it is important to control this variation. This chapter develops techniques or scenarios to understand the source of harmonics following several scenarios by combining linear and non-linear loads. In addition, this chapter will discuss methods of measuring harmonics using electrical analyzers.

Keywords: power quality, nonlinear loads, electrical analyzer, total harmonic distortion, impact of harmonics

1. Introduction

Today, electrical energy has become the pedestal of socio-economic development in every country. However, the rate of electricity access in sub-Saharan Africa is low, at 50.6% in 2021 [1]. In the context of technological expansion, the impact of climate change and the energy transition, it is necessary to increase the coverage rate in the electricity field. This has allowed to technological evolution in the power system chain. In other words, the conventional power system, which consists of generating, transmitting and distributing has modernized (smart grid) [2]. The smart grid is a dynamic, stable, flexible, resilient and controllable system.

Mainly with the management function via programmable controllers, the hybrid systems have become the optimal solutions for meeting electrical energy demand efficiently [3]. So, the redeployment of renewable energies in electricity production ensures a smooth energy transition and reduces the rate of CO_2 emissions [4].

In addition, the electronics integration into electrical systems and the developments of electrical technology and/or electronic devices have allowed the proliferation of non-linear loads consuming little electrical energy from the users of electrical energy. Electronic converters and non-linear loads generate disturbances in the network such as an increase in the rate of harmonics, current unbalance and consumption of reactive energy, etc. The use of electronic devices in an electrical installation impacts the quality of the electrical energy of the grid and the proper functioning of the devices [5]. The permanent presence of harmonics has harmful technical and

IntechOpen

economic consequences. It distorts the quality of the energy distributed and reduces the lifespan of the loads and the elements of the distribution grid, which allow the transmission of energy to the customers [6].

However, to ensure the quality of electrical energy in accordance with EN 50160 and IEEE 1159–2009 standards, passive and active techniques must be developed [7].

The first technique (passive) is detailed in this document. It consists of having historical electrical data from the electrical grid. This involves taking measurements with accurate electrical devices. Electrical analyzers are the most suitable for studying the quality of electrical energy. Indeed, they make it possible to record in real-time all the electrical quantities and the electrical disturbances of the electrical systems. In this chapter, it is also a question of approaching the analysis of the mitigation and the sources of electrical disturbances and their mode of propagation.

2. Definition of linear and non-linear loads

According to EN 50160, an electrical harmonic is a component of sinusoidal signal voltage or current and frequency multiple of the fundamental frequency (50 Hz).

2.1 Linear load

A linear load in electricity is any load that draws a current of the same shape as the voltage, in other words quasi-sinusoidal. For example, for a resistive load, we find the representation of the current and the tension which are all of identical form in **Figure 1**.

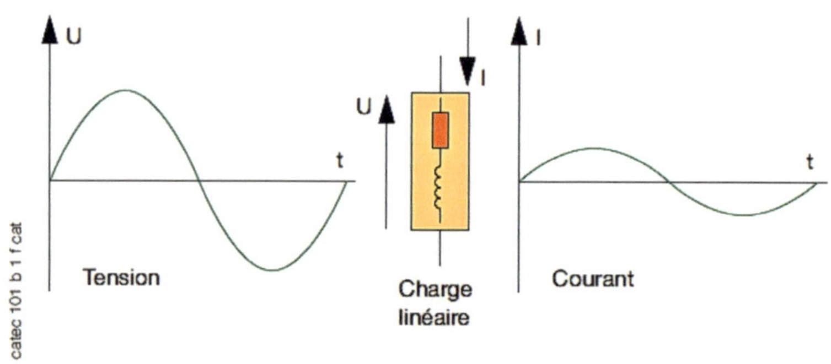

Figure 1.
Example of a waveform of the electrical signal of a linear load [8].

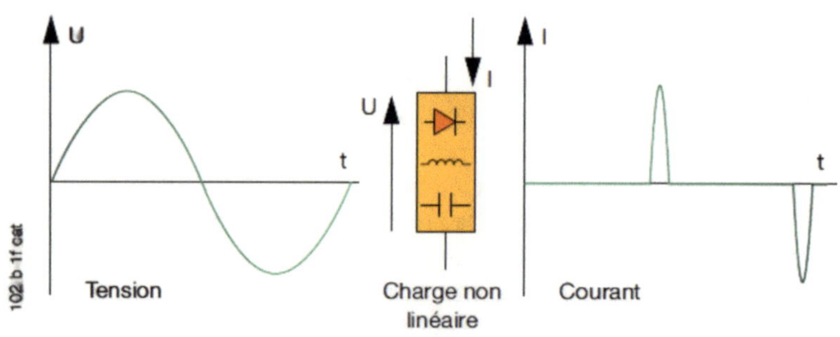

Figure 2.
Example of the waveform of the electrical signal of a nonlinear load [8].

2.2 Nonlinear load

A nonlinear load is a load that draws a current that distorts the shape of the voltage. With a variable speed drive (as a load), we find the behavior or forms of the electrical magnitude as shown in **Figure 2**.

3. Fundamental indicators of electrical harmonics

The study of harmonics is based on signal theory. However, mathematical tools are essential in signal processing. Harmonics are defined as sinusoidal signals of multiple frequencies at the considered fundamental frequency. Thus, Fourier's theorem is one of the methods that allows development of any periodic signal of distorted waveform into sums of sinusoidal signals of different amplitudes.

In the field of physics, more particularly in industrial applications such as the distribution of electrical energy, measurement instruments have been designed to measure and analyze the level of harmonic disturbance [7, 9, 10].

In fact, several approaches have been developed that take into account the harmonic part. Among these, that of Baudeau takes into account the presence of harmonics [11].

The active and reactive powers are expressed respectively by the Eqs. (1) and (2):

$$P = \Sigma_n V_n * I_n * \cos\phi_n \tag{1}$$

$$Q = \Sigma_n V_n * I_n * \sin\phi_n \tag{2}$$

With n: whole number defining the harmonic order.

Thus, the design power or apparent power is expressed by the following formula:

$$S^2 = P^2 + Q^2 + D^2 \tag{3}$$

With, D: distorting power (VAD), P: active power and Q: reactive power.

3.1 Power factor

An electrical parameter defining the level of quality of the power consumed is the power factor. It is used by the equation:

$$FP = \frac{P}{S} \tag{4}$$

With, P: active power, S: total apparent power (fundamental + harmonic components).

3.2 Order of harmonics

The order of the harmonics is the ratio of the frequency of the harmonic (fn) to the frequency of the referential or fundamental sinusoidal function ($f1$). It is written as follows [7]:

$$N = \frac{f_n}{f_1} \tag{5}$$

3.3 Effective value of the harmonic quantity

For a deformed magnitude and, in a steady state, the energy dissipated by the Joule effect is the sum of the energies dissipated by each of the harmonic components [7], i.e.:

The effective value of the harmonic magnitude.

$$\mathrm{Y} = \sqrt{\sum_{n=1}^{n=\infty}\left(Y_n^{\,2}\right)} \tag{6}$$

3.4 Total harmonic distortion (THD)

This quantity allows us to detect grids polluted with harmonics, it can be evaluated with the harmonic distortion rate, according to the definition given by the IEC dictionary: this parameter, is also called harmonic distortion or distortion factor. The harmonic content can be calculated either with [9, 10]:

The global harmonic rate of distortion. It represents the ratio of the effective value of the harmonics *Yn* to the effective value of the fundamental *Y*1 [12, 13].

$$THD(\%) = 100\frac{\sqrt{\sum_{n=2}^{n=\infty} Y_n^{\,2}}}{Y_1} \tag{7}$$

Where, *Yn*: effective value of (*n*-1) harmonics and *Y*1: effective value of the fundamental.

It is possible to calculate the individual distortion rate by using the Eq. (8) [7]:

$$THI(\%) = 100\frac{Y_n}{Y_1} \tag{8}$$

The Eq. (9) is used to evaluate the THD in electrical network distribution fixed by the IEEE 519 standard provides guidelines for harmonic current limits at the point of common coupling (PCC) [14, 15].

$$THD(\%) = 100\frac{\sqrt{\sum_{n=2}^{n=\infty} I_n^{\,2}}}{I_L} \tag{9}$$

I_n: the maximum rms value delivered in the fundamental state.

I_L: Current loads connected to the PCC.

4. Behavior of harmonics

Several indicators define the presence of harmonics, they make it possible to evaluate the quality of the electrical energy and also specify the types of distortions present. The C.A 8220 measuring device is the power analyzer used to measure the harmonic disturbances likely to be injected into the electrical distribution networks via electrical loads.

To study the behavior of the harmonic rate in the presence of electrical charges, several scenarios have been developed (**Table 1**).

The principle consists of connecting the analyzer to the load terminals (see **Figure 3**) to record the harmonic rate of the loads summarized in **Table 2**.

In single-phase mode, the analyzer is connected in this experiment as follows (**Figure 4**).

The results presented in this part, are to evaluate the harmonic rate of some devices and then to observe the behavior of the content of the harmonics according to the scenarios:

- The combination of moderately polluting loads (a compressor and an air conditioner);
- The combination of highly polluting loads (computers and ultrasound machines) and a pure load (water heater).

The figure below gives some illustrations of the measurement devices and the principle of data collection (**Figure 5**).

Name of Loads	Power of Loads
Laptop	65 W
Ultrasound-machine	400 W
Air Compressor	750 W
Air Conditioner	1980 W
Water-heater	2200 W
Printer	365 W
Photocopier	1850 W
Fridge	75 W
TV	55 W
Energy-saving light bulb	9 W
Fluorescent Bulb	85 W
Incandescent Bulb	60 W
Cellphone charger	10 W

Table 1.
Power of loads used.

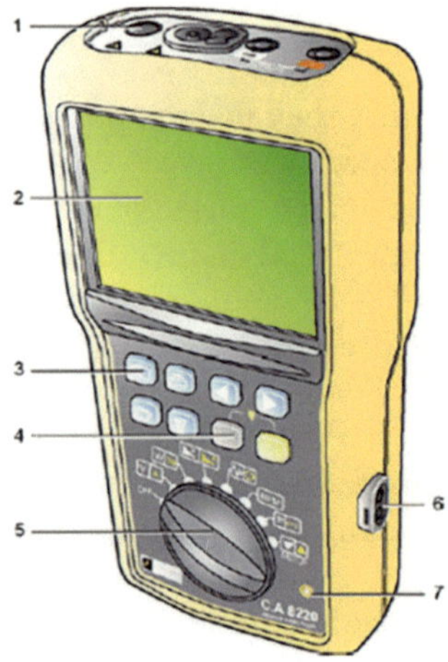

Rep.	Fonction
1.	Bornes électriques.
2.	Ecran LCD avec rétroéclairage.
3.	Touches bleues.
4.	Touches blanche/jaune.
5.	Commutateur rotatif.
6.	Interface optique infrarouge.
7.	Témoin lumineux d'alimentation externe.

Figure 3.
CA8220 analyzer overview [16].

5. Results obtained

Based on the measurement principle explained above, it appears that the level of distortion varies according to the technology of the electrical charges. The table below summarizes the level of harmonic distortion recorded at the terminals of the targeted electrical devices.

5.1 Combination: water heater, air conditioner, compressor, ultrasound machine and computer

However, by connecting the combination of these devices to a low-voltage power supply (220 V, 50 Hz). As shown in **Figure 6** we observe a variation of the harmonic rate.

5.2 Summary of results

Tables 2 and **3** below summarize respectively the global harmonic rate of each device and the variation of the harmonic rate according to the combinations of these devices.

5.3 Analysis and interpretation of results

The first scenario begins with the powering up of a computer, we obviously see that the harmonic rate is high, estimated at 194.5%. This type of load can then be classified as a very polluting load. Once this computer and an ultrasound scanner are supplied at the same time, this combination leads to a drop in the Total Harmonic Distortion (THD) to 64.9%, thus the combination of two polluting loads of different contents leads to a significant drop in the THD. This time, in scenario 3, it is a case of supplying the devices of scenario 2 and a resistive load (the water heater) at the same time, the result obtained is very interesting because the harmonic rate is low with a THD of 4.9%.

Electrical loads	Harmonic ranks	THD (%)
Fluorescent lamp	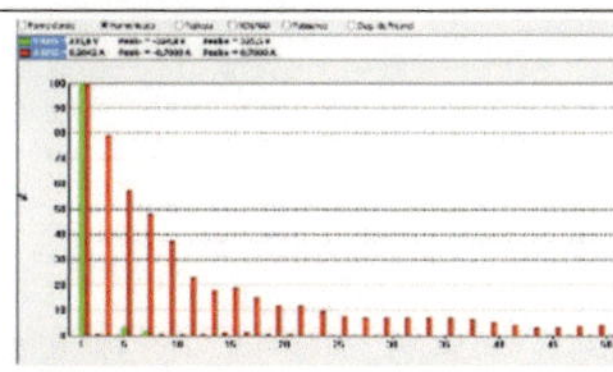	125.1
Ballast lamp	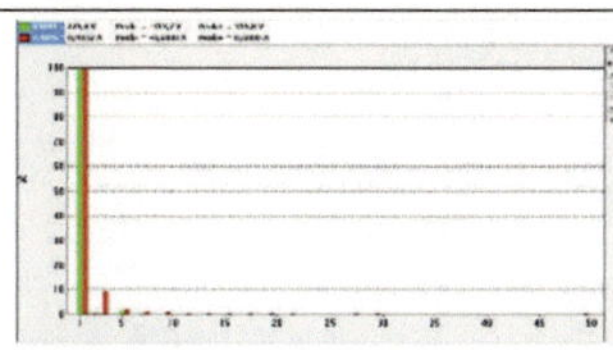	10
Incandescent lamp	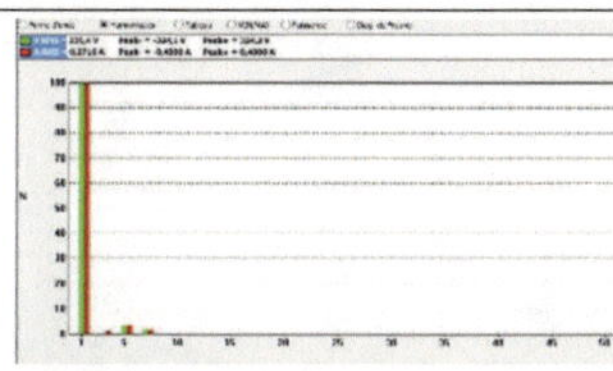	2.5
Fan	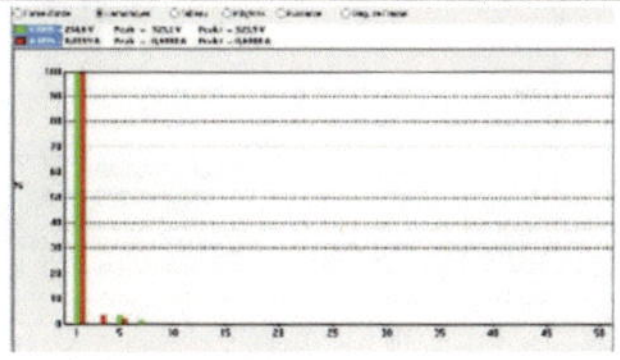	4.1
Flat screen TV	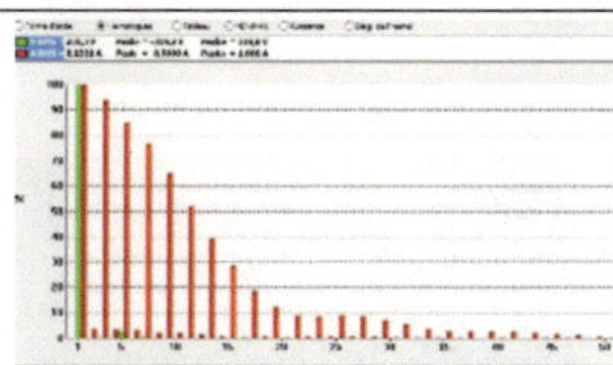	179.8
Phone charger	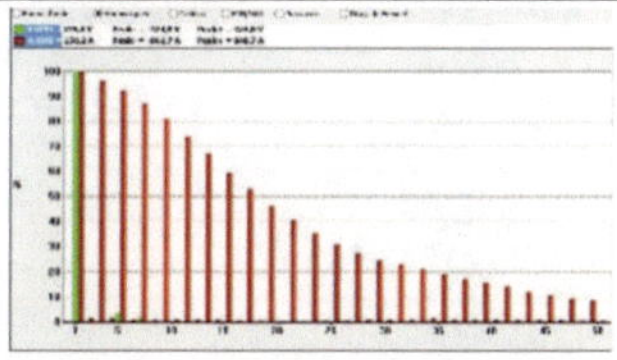	241.9
Printer	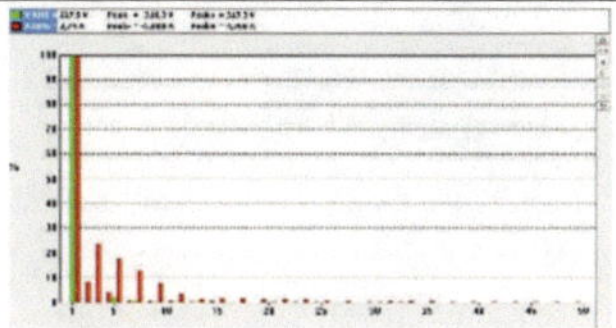	21
Air conditioner	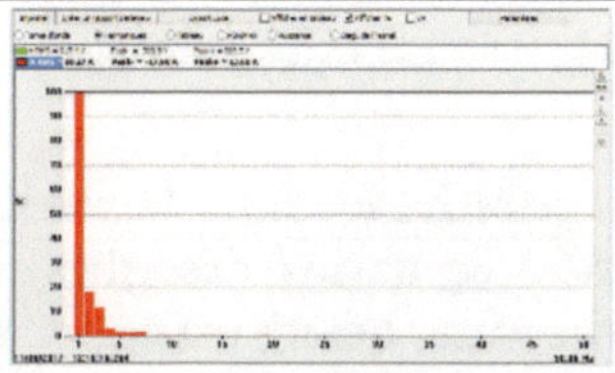	21.1

Electrical loads	Harmonic ranks	THD (%)
Computer	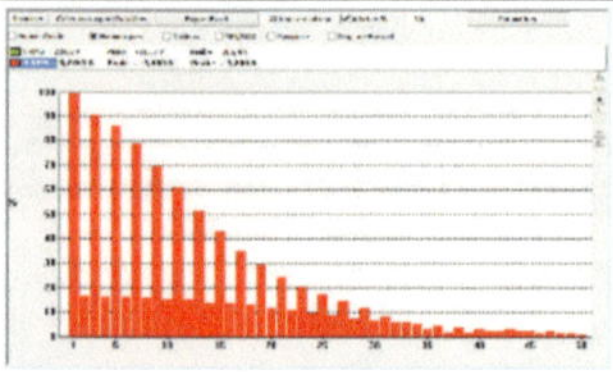	197.9

Table 2.
Some results obtained at the terminals of the targeted loads.

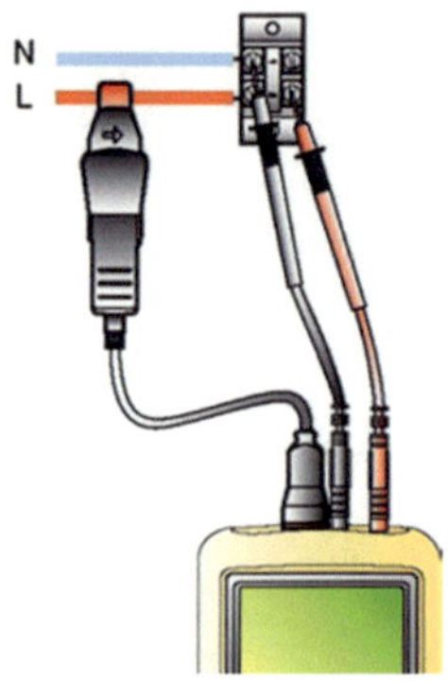

Figure 4.
Single-phase connection [16].

Figure 5.
Mesure des harmoniques électriques générés par les appareils monophasés.

This explains the impact of resistive loads in the attenuation of harmonic pollution. As demonstrated in scenario 3, scenario 6 describes a consequent attenuation of the harmonic rate also due to the water heater in the presence of non-linear loads (the air conditioner and the compressor), both of which consume energy. However, when the water heater is removed from the system, a rise in the THD is observed.

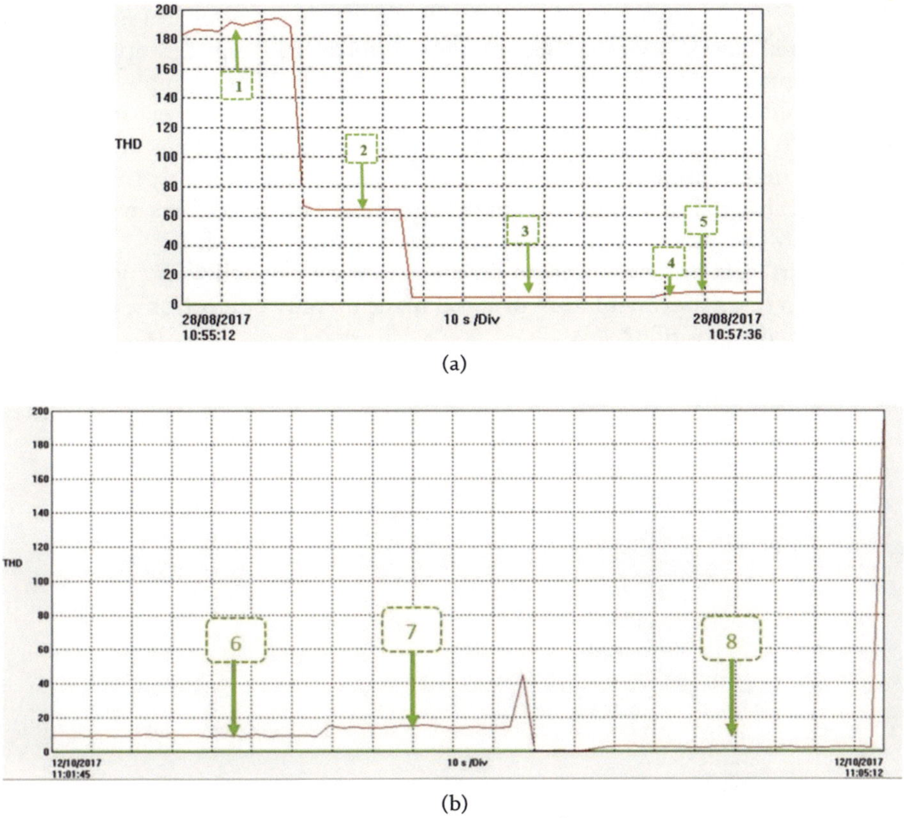

Figure 6.
Variation of the harmonic rate produced by the combination of linear and non-linear loads.

CODE	Loads	THDA
1	Computer	194.5
2	Computer + Ultrasound	64.9
3	Computer + Ultrasound + Heater	4.9
4	Computer + Ultrasound + Heater + Compressor	5.2
5	Computer + Ultrasound + Heater + Compressor + Air-conditioner	8.4
6	Heater + Compressor + Air condition	7.9
7	Compressor + Air condition	13.44
8	Computer + Heater	2.82

Table 3.
Variation of harmonic rate.

6. Conclusion

The experimentation carried out in this chapter has made it possible to observe the variation in the harmonic rate resulting from the use of nonlinear and linear loads. Eight (8) combinations of loads called "Scenario" have been developed in order to clearly identify the propagation of electrical harmonics in an electrical network. The results obtained have shown that the presence of linear loads significantly attenuates harmonic disturbances although they are energy-intensive. For example, turning on a computer (nonlinear load) and a water heater (linear load) drew a very low harmonic current (2.8%) compared to the THD of the computer (197.9%). Thus, linear loads contribute to maintaining the quality of electrical energy, particularly the waveform of the electrical signal that must exist in the equipment of subscribers even if the new WAEMU(UEMOA) energy efficiency policy tends to be replaced.

In perspective, in order to improve the energy efficiency policy, would it be necessary to deepen this experimentation in a grid made up only of electronic devices with low electrical consumption?

Author details

Abdourahimoun Daouda
Abdou Moumouni University, Niamey, Niger

*Address all correspondence to: daoudaabdourahimoun@gmail.com

References

[1] United Nations Conference on Trade and Development, Commodities at a Glance: Special Issue on Access to Energy in Sub-saharan Africa. In: Commodities at a Glance. United Nations. 2023. DOI: 10.18356/9789210024808

[2] Idda A, Said B, et al. Hybridation d'une Centrale Diesel en Energie Photovoltaïque. Algerie: Université Ahmed Draïa-Adrar; 2013. Thèse de doctorat

[3] Momoh JA. Smart Grid: Fundamentals of Design and Analysis. Canada: IEEE; 2012

[4] Transitions to low carbon electricity systems: Key economic and investments trends Changing course in a post-pandemic world [Online]. Available from: https://www.iaea.org/sites/default/files/21/06/transitions-to-low-carbon-electricity-systems-changing-course-in-a-post-pandemic-world.pdf [Accessed: December 16, 2023]

[5] Farooq H, Zhou C, Allan M, et al. Investigating the power quality of an electrical distribution system stressed by non-linear domestic appliances. Renewable Energy and Power Quality Journal. 2011;**1**(9):283-288

[6] Alali, Mohamad Alaa Eddin. Contribution à l'étude des compensateurs actifs des réseaux électriques basse tension. 2002. Available from: https://hal.science/hal-00722521

[7] Abdourahimoun D, Saidou M. Étude des perturbations harmoniques et leurs traitements dans le réseau électrique: cas du réseau BT de la ville de Niamey. London: EUE; 2018

[8] Socomec. Cahier technique: systèmes de coupure et de protection. 2011. 120 p

[9] Schlabbach J, Blume D, Stephanblome T. Voltage Quality in Electrical Power Systems. London: United Kingdom; 1999

[10] Shmilovitz D. On the definition of total harmonic distortion and its effect on measurement interpretation. IEEE Transactions on Power Delivery. 2005;**20**(1):526-528

[11] Bogale, Wessen S. Evaluating the Level of Harmonic Distortion in a Typical Distribution Feeder. 2015. Available from: https://digitalscholarship.unlv.edu/thesesdissertations/2333/

[12] Collombet C, Lupin JM, Schonek J. Harmonic disturbances in networks, and their treatment. Cahier Technique Merlin Gerin. 1999;**1**(152):1-31

[13] La Rosa D, Harmonics F. Power systems. In: Electric Power Engineering Series. 2006. Available from: http://103.62.146.201:8081/jspui/handle/1/4273

[14] Sugano Y, Keetels M, Vroomen J. Audio-motor but not visuo-motor temporal recalibration speeds up sensory processing. PLoS One. 2017;**12**(12):e0189242

[15] Sankaran C. Power Quality. USA: CRC Press; 2017

[16] Manuel d'utilisation de C.A 8220

Chapter 3

Perspective Chapter: Frequency Domain Models of Nonlinear Loads in Power Systems

Florin Constantinescu, Alexandru Gabriel Gheorghe, Mihai Eugen Marin and Florin Roman Enache

Abstract

Section 1 deals with harmonic current sources models for one diode and two diodes rectifiers, the model parameters being determined by simulations. Section 2 describes frequency domain (FD) models of nonlinear home appliances whose parameters are determined by measurements. Some FD models for the loads using firing angle control devices are presented in Section 3. The fourth section presents the simulations of a relatively large circuit in the time domain (TD) and in the FD using the algorithms of CADENCE and ADS. The HB analysis of ADS using the proposed FD models proves to be the most efficient for this example.

Keywords: nonlinear devices, home appliances, frequency domain models, harmonic balance, frequency domain analysis, current source models

1. Introduction

Although AC networks contain only undistorted sinusoidal sources of the same frequency f, their actual voltages and currents often have other spectral components including several odd and sometimes even multiples of f. The presence of these spectral components (called harmonics) is due to nonlinear loads such as arcing devices, saturated transformers, and various types of rectifiers that power electric transport vehicles or electronic equipment.

The additional losses caused by current harmonics can be reduced by complementary power compensation. In order to bring the harmonic pollution in the power grid to an acceptable level [1], some iterative algorithms are used to design the compensation circuit or to determine its optimal size and location to minimize the effects of harmonics [2]. This type of calculation sometimes uses evolutionary algorithms, the efficiency of which depends on the effectiveness of the variant analysis of nonlinear energy systems.

Usually, periodic steady-state computation of nonlinear power systems is performed by the time domain analysis. Some transient analysis programs such as EMTP (Electromagnetic Transient Program) or EMTDC (Electromagnetic Transient Design and Control) use the "brute force" approach (integrating circuit equations

IntechOpen

until all transients decay) to achieve this [3, 4]. This method is not efficient for circuits with time constants much larger than the signal period. In this case, is preferred the shooting method using the Newton-Raphson algorithm [4].

Another method for calculating the periodic steady state of nonlinear circuits is the frequency domain analysis. The best method of this kind is the harmonic balance (HB) algorithm, which works on both frequency domain and time domain representations of the same signal [4, 5]. A signal is represented in the time domain by a set of evenly spaced samples, while in the frequency domain by a set of spectral components. These two sets are related by the direct discrete Fourier transform and inverse discrete Fourier transform, so the results of time domain analysis can be easily transformed to the frequency domain, and the results of frequency domain analysis can also be transformed to the time domain [5].

2. Current sources models with parameters determined by simulations

2.1 One diode rectifier

The circuit in **Figure 1** is driven by a sinusoidal voltage source with amplitude V = 220 V and f = 50 Hz, having C_1 = 2 mF and R_1 = 37 Ω. The rectifier is connected to the AC source by a cable of 10mm^2, with a length of 30 m (R_{line} = 0.1104 Ω, L_{line} = 72 μH).

Starting from the linear dependence of the current harmonics amplitudes and phases on the amplitude and phase on the input voltage fundamental component suggested in Ref. [3] and using the simplified rectifier schematic described in Refs. [6, 7], the model of the frequency-defined device (FDD) in **Figure 2** has been built. This model is described by a linear-controlled current source for each harmonic component, the control parameters being the amplitude and the phase of the voltage fundamental component V_1.

This model is given in **Figure 2**, followed by its ADS description [8].

V_1 = V(out1)

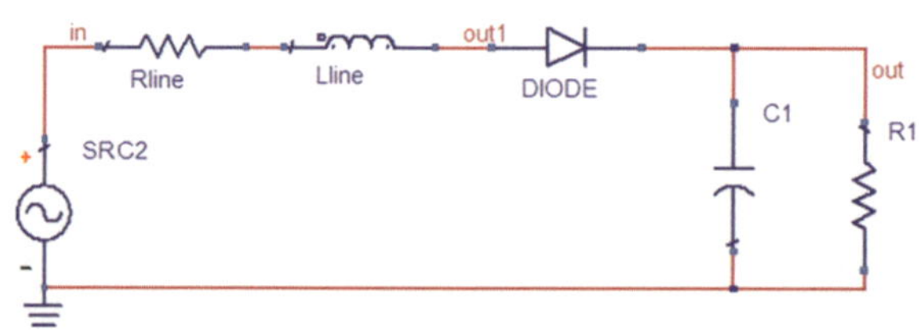

Figure 1.
One diode rectifier with a C filter.

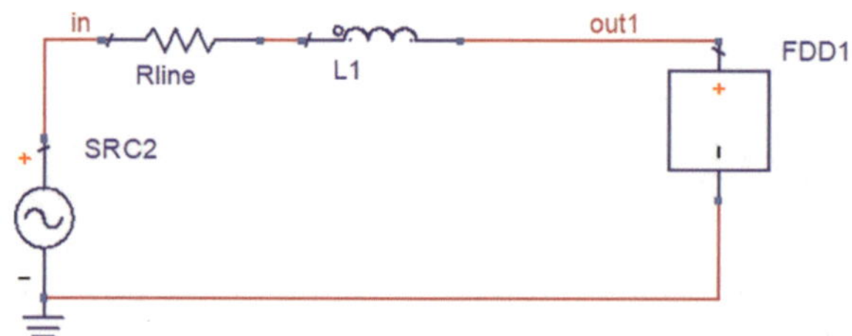

Figure 2.
Linear-controlled source model for the circuit in ***Figure 1****.*

I_0 = polar(1.7e-2*mag(V_1), 0)
I_1 = polar(3.14e-2*mag(V_1), 5 + 1*phase(V_1))
I_2 = polar(2.45e-2*mag(V_1), 11 + 2*phase(V_1))
I_3 = polar(1.54e-2*mag(V_1), 17 + 3*phase(V_1))
I_4 = polar(6.72e-3*mag(V_1), 25 + 4*phase(V_1))
I_5 = polar(6.83e-4*mag(V_1), 73 + 5*phase(V_1))
I_6 = polar(2.48e-3*mag(V_1), −158 + 6*phase(V_1))
I_7 = polar(2.53e-3*mag(V_1), −147 + 7*phase(V_1))
I_8 = polar(1.07e-3*mag(V_1), −134 + 8*phase(V_1))
I_9 = polar(5.059e-4*mag(V_1), 21 + 9*phase(V_1))
I_{10} = polar(1.145e-3*mag(V_1), 41 + 9*phase(V_1))

The coefficients in the amplitude and phase description of the current harmonics are identified by simulations of the circuit in **Figure 1**.

A test circuit consisting of 10 identical one-diode rectifiers, connected to a sinusoidal source with f = 50 Hz and amplitude V = 220 V through a 50 mm^2 cable of 30 m (**Figure 3**) has been used for the harmonic balance analysis, where R_{line1} = 34.8 mΩ and L_{line1} = 57.6 μH.

2.2 Two diode rectifier

A similar model can be made for a two diode rectifier (**Figure 4**). This circuit represents a reduced model of a compact fluorescent lamp [6, 7]. The parameter values are $C_1 = C_2$ = 15 μF, R_1 = 5 kΩ.

In this case, the line is a 1 mm^2 cable with a length of 30 m, Rline = 1.08 Ω, and Lline = 72 μH. This model is described as follows [8]:

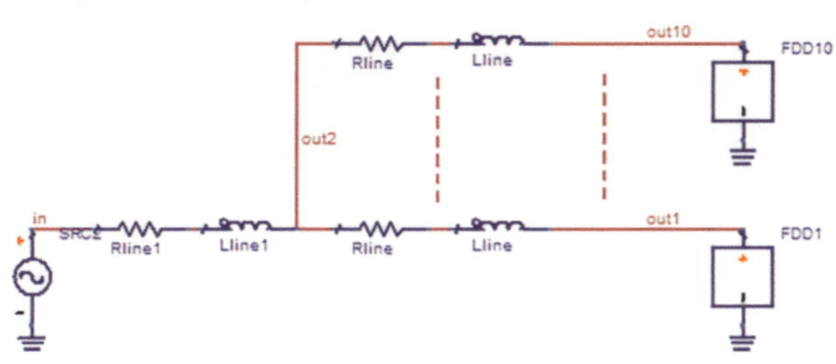

Figure 3.
One-phase test circuit.

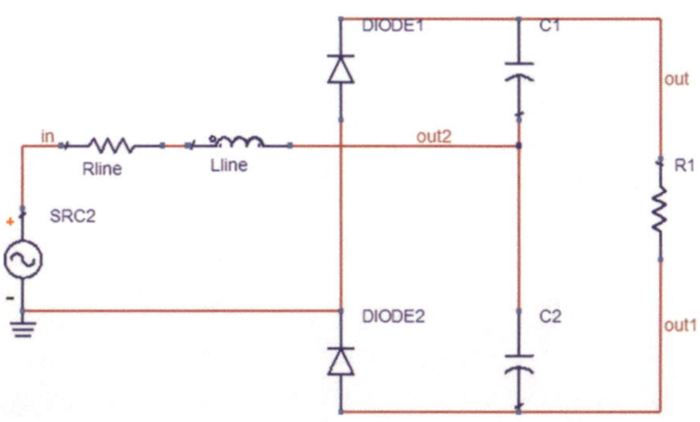

Figure 4.
Two diode rectifier.

V_1 = V(out2)
I_1 = polar(1.276e-3*mag(V_1), 31 + 1*phase(V_1))
I_3 = polar(1.024e-3*mag(V_1), 94 + 3*phase(V_1))
I_5 = polar(6.453e-4*mag(V_1), 164 + 5*phase(V_1))
I_7 = polar(3.394e-4*mag(V_1), −105 + 7*phase(V_1))
I_9 = polar(2.643e-4*mag(V_1), 9 + 9*phase(V_1))
I_{11} = polar(2.501e-4*mag(V_1), 101 + 11*phase(V_1))
I_{13} = polar(1.924e-4*mag(V_1), −165 + 13*phase(V_1))
I_{15} = polar(1.5961e-4*mag(V_1), −60 + 15*phase(V_1))
I_{17} = polar(1.532e-4*mag(V_1), 38 + 17*phase(V_1))
I_{19} = polar(1.3174e-4*mag(V_1), 133 + 19*phase(V_1))
I_{21} = polar(1.161e-4*mag(V_1), −124 + 21*phase(V_1))

The test circuit similar to the one in **Figure 3**, where 10 identical two diode rectifiers are connected to a sinusoidal source with f = 50 Hz and amplitude V = 220 V through a 10 mm^2 cable of 30 m has been used for the harmonic balance analysis, where R_{line1} = 110.4 mΩ, L_{line1} = 72 μH.

2.3 Analysis

The test circuit of 10 identical one-diode rectifiers, connected to a sinusoidal source of f = 50 Hz and amplitude V = 220 V through a 50 mm^2 cable of 30 m with R_{line1} = 0.0348 Ω, L_{line1} = 57.6 μH, (**Figure 5**) has been used for the harmonic balance analysis [6].

This section presents the results of a harmonic balance analysis performed using ADS. Two circuits with the structure shown in **Figure 5** are analyzed. The current

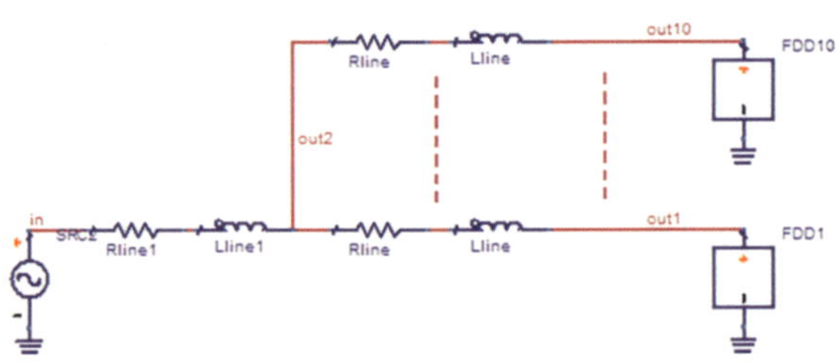

Figure 5.
One-phase test circuit.

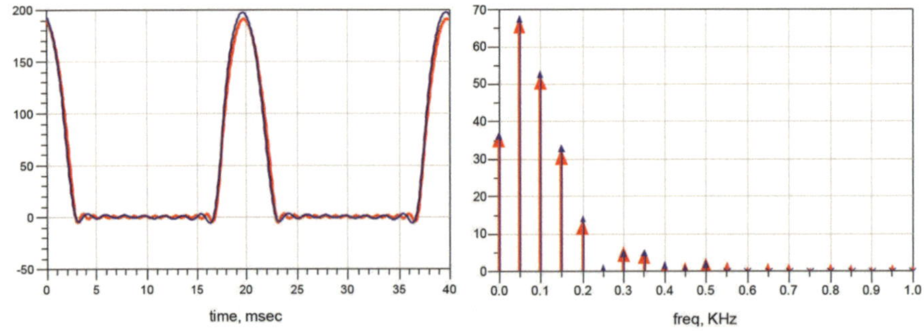

a) waveform b) spectrum

Figure 6.
Source current for the circuit in ***Figure 5*** *with one diode rectifier: (a) waveform and (b) spectrum.*

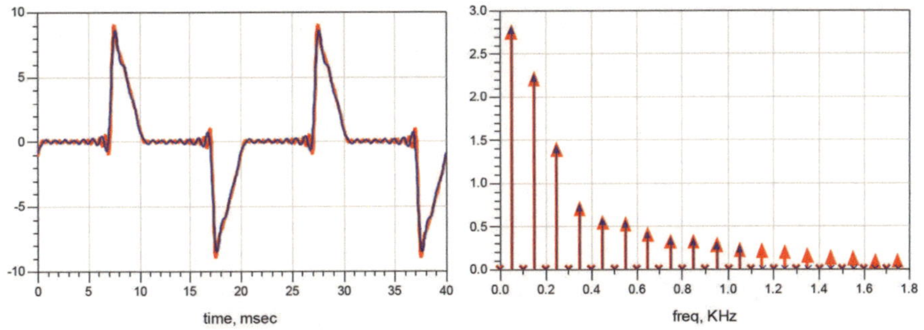

a) waveform b) spectrum

Figure 7.
*Source current waveform for the circuit in **Figure** 5 with two diode rectifiers: (a) waveform and (b) spectrum.*

through the sinusoidal voltage source can be considered as a global variable representing the behavior of the test circuit. Results shown as red waveforms or spectral lines are obtained using a classical nonlinear model defined by time domain eqs. (TD model). Results shown as blue waveforms or spectral lines were determined using the models described in Sections 1.1 and 1.2.

The current through the sinusoidal voltage source can be viewed as a global variable representing the behavior of the test circuit (**Figure 6**) [8].

The results obtained using the classical time-domain model are similar to those obtained using the proposed model, in the both time and frequency domains. The analysis using the classical time-domain model takes 0.96 seconds, while the analysis using the proposed model takes only 0.17 seconds. In both cases, 20 significant harmonic components are considered.

The harmonic balance analysis of a test circuit of 10 two-diode rectifiers produced similar results (**Figure** 7). In this case, the simulation time is 4.12 seconds using the classical nonlinear model described in the time domain, but only 0.19 seconds using the proposed model.

These simulations have been performed with 35 harmonic components. It is obvious that as the harmonics number increases, a greater simulation time is obtained using FD models.

Some other properties of these FD models are described in Ref. [8]: It can reproduce the third current harmonic reduction as the length of a line supplying fluorescent lamps increases [6]. The accuracy of this simple model is not preserved for long lines of about several Km.

The proposed FD model reduces the simulation time by about an order of magnitude compared to the classical TD model using nonlinear loads described in the time domain. This result can be explained as follows:

- using the proposed model, the harmonic balance works only in the frequency domain, whereas using the classical model with nonlinear components, the algorithm works in both the frequency and time domains,
- no source stepping is required when using the proposed model, unlike when using the TD model in the same circuit.

The content of the paragraph "1. Current sources models with parameters determined by simulations" reproduces the research reported for the first time in Ref. [8].

3. Current source models with parameters determined by measurements

Numerous home appliances have complex electronic circuits, and in some cases, circuit diagrams and/or unknown parameters are not available and the methods in Section 1 cannot be used. A new approach to computing the parameters of their FD models, based on measurements, is described in this section [9, 10].

An analyzer is a relatively simple apparatus able to compute the modules and phases of the harmonic components of a non-sinusoidal waveform; the results are given with two significant digits. For example, an air conditioning system whose voltage and current waveforms are given in **Figure 8** has its measured current harmonic components given in **Table 1**. Considering these harmonic components, the waveform in **Figure 9** is rebuilt. This waveform is, at first glance, very close to that in **Figure 8**. Similar FD models have been built for a vacuum cleaner, a microwave oven,

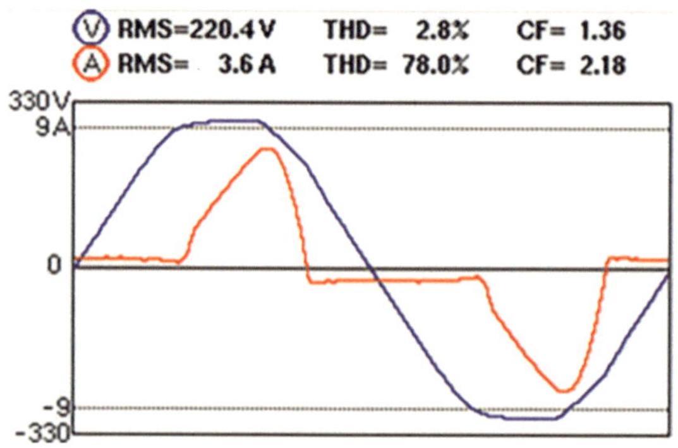

Figure 8.
Air conditioning system voltage and current waveforms—Measurements.

H order	1	3	5	7	9	11	13	15	17	19
I [A]	2.8	1.9	0.75	0.3	0.3	0.1	0.1	0.1	0.1	0.1
Phase [°]	0	147	−70	2	126	−120	−80	55	110	−140

Table 1.
Air conditioning system measured current harmonic components.

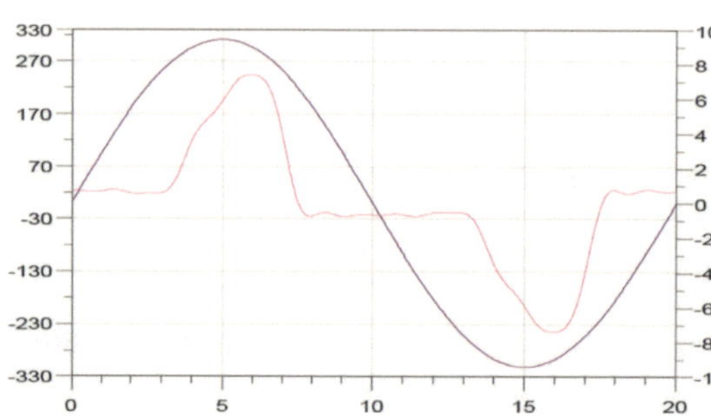

Figure 9.
Air conditioning system voltage and current waveforms—Simulation.

and a compact fluorescent lamp [8]. A three-phase network containing the FD models mentioned before has been simulated with the usual TD models and with proposed FD models. Because of the low precision of the measurements made with the analyzer (1–3 digits), the results obtained with these FD models are not accurate.

A network analyzer is a more sophisticated device which measures a set of several hundred time—samples to characterize a current waveform. A program for the computation of the modules and phases of the harmonic components delivers accurate values of these magnitudes. The results of this kind of measurement and computations are described in Ref. [10]. The FD models are given as results of linear interpolations of the measured data in the format proposed in Ref. [3]. Some results are given below for several devices (**Figures 10** and **11**, **Tables 2–8**).

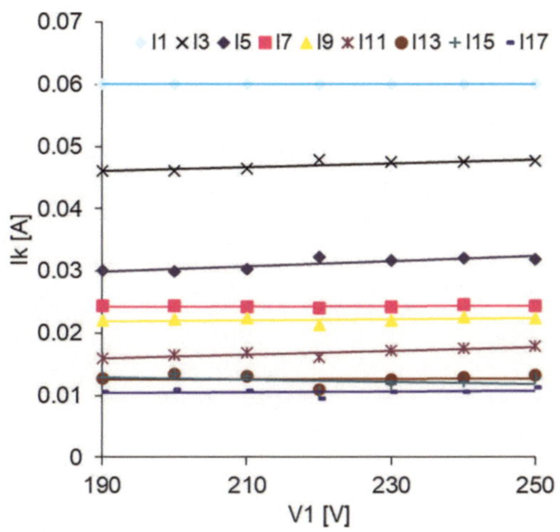

Figure 10.
RMS values of the current harmonics for a CFL 10 W.

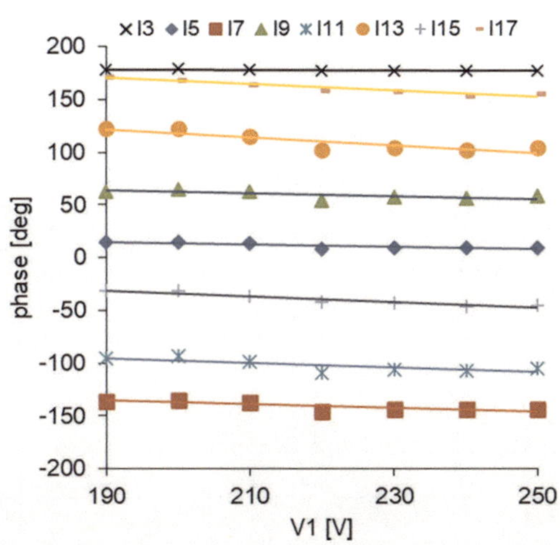

Figure 11.
Phases of the current harmonics for a CFL 10 W.

RMS [A]	PHASE [°]
rms(I1) = −5E-18 V1 + 6E-2	ph(I1) = 0
rms(I3) = 3E-5 V1 + 4E-2	ph(I3) = −2.8E-2 Φ1 + 184
rms(I5) = 4E-5 V1 + 2.2E-2	ph(I5) = −0.11 Φ1 + 35
rms(I7) = 1E-6 V1 + 2.4E-2	ph(I7) = −0.17 Φ1-102
rms(I9) = 7E-6 V1 + 2.1E-2	ph(I9) = −0.14 Φ1 + 90
rms(I11) = 3E-5 V1 + 1E-2	ph(I11) = −0.21 Φ1-55
rms(I13) = 2E-7 V1 + 1.3E-2	ph(I13) = −0.38 Φ1 + 193
rms(I15) = −2E-5 V1 + 1.6E-2	ph(I15) = −0.27 Φ1 + 20
rms(I17) = 6E-6 V1 + 9.3E-3	ph(I17) = −0.3 Φ1 + 228

Table 2.
TFD model for a CFL 10 W.

The FD models of other nonlinear loads containing diode rectifiers are given below [10].

RMS [A] for LED 5 W	PHASE [°] for LED 5 W
rms(I1) = 0.2	ph(I1) = 0
rms(I3) = 2E-5 V1 + 1.3E-2	ph(I3) = −0.06 Φ1 + 186
rms(I5) = 4E-5 V1 + 4.6E-3	ph(I5) = −0.17 Φ1 + 33
rms(I7) = 3E-5 V1 + 2.7 E-3	ph(I7) = −0.39 Φ1-83
rms(I9) = 1E-5 V1 + 5.3E-3	ph(I9) = −0.49 Φ1 + 136
rms(I11) = 2E-5 V1 + 3.2E-3	ph(I11) = −0.48 Φ1-33
rms(I13) = 3E-5 V1-8.2E-4	ph(I13) = −0.64 Φ1 + 197
rms(I15) = 2E-5 V1-2.8E-4	ph(I15) = −0.95 Φ1 + 106
rms(I17) = 2E-5 V1-1.1E-3	ph(I17) = −0.98 Φ1 + 315

Table 3.
FD model for LED 5 W.

RMS [A] for LED 17 W	PHASE [°] for LED 17 W
rms(I1) = −3.9E-4 V1 + 1.6E-1	ph(I1) = 0
rms(I3) = −2.3E-4 V1 + 1.2E-1	ph(I3) = −3.9E-2 Φ1 + 185
rms(I5) = −5E-5 V1 + 6 E-2	ph(I5) = −0.17 Φ1 + 38
rms(I7) = −2E-5 V1 + 4E-2	ph(I7) = −0.44 Φ1-67
rms(I9) = −8E-5 V1 + 4.6E-2	ph(I9) = −0.54 Φ1 + 157
rms(I11) = −3E-5 V1 + 3.2E-2	ph(I11) = −0.55 Φ1-5
rms(I13) = 4E-5 V1 + 1.1E-2	ph(I13) = −0.77 Φ1 + 241
rms(I15) = 7E-6 V1 + 1.4E-2	ph(I15) = −1.11 Φ1 + 157
rms(I17) = 2E-5 V1 + 6.5E-3	ph(I17) = −1.12 Φ1+ 362

Table 4.
FD model for LED 17 W.

RMS [A] for air conditioner (heating)	Phase [°] for air conditioner (heating)
rms(I1) = −7E-3 V1 + 6.14	ph(I1) = 0
rms(I3) = −1.6E-3 V1 + 1.21	ph(I3) = −0.1 Φ1 + 142
rms(I5) = 7E-4 V1 + 0.47	ph(I5) = −0.02 Φ1-92
rms(I7) = 4E-4 V1 + 0.16	ph(I7) = 0.14 Φ1-34
rms(I9) = −8E-4 V1 + 0.32	ph(I9) = −0.21 Φ1 + 118
rms(I11) = 8E-4 V1 + 1.6E-2	ph(I11) = −0.18 Φ1-122
rms(I13) = 5E-4 V1 - 5E-2	ph(I13) = 0.88 Φ1-245
rms(I15) = −7E-4 V1 + 0.21	ph(I15) = −0.28 Φ1 + 52
rms(I17) = 5E-4 V1-4.5E-2	ph(I17) = −0.02 Φ1 + 108

Table 5.
FD model for air conditioner (heating).

RMS [A] for air conditioner (cooling)	PHASE [°] for air conditioner (cooling)
rms(I1) = −5E-3 V1 + 3.38	ph(I1) = 0
rms(I3) = −1.5E-3 V1 + 1.94	ph(I3) = −0.24 Φ1 + 192
rms(I5) = 1.8E-3 V1 + 0.32	ph(I5) = −0.29 Φ1-19
rms(I7) = −1.2E-3 V1 + 0.49	ph(I7) = 0.01 Φ1-15
rms(I9) = −5E-05 V1 + 0.31	ph(I9) = −0.66 Φ1 + 246
rms(I11) = 1.9E-3 V1-0.29	ph(I11) = −0.65 Φ1 + 17
rms(I13) = −1.4E-3 V1 + 0.39	ph(I13) = −0.86 Φ1 + 63
rms(I15) = 4E-4 V1-2E-4	ph(I15) = −1.02 Φ1 + 248
rms(I17) = −3E-4 V1 + 8.7E-2	ph(I17) = 0.88 Φ1-113

Table 6.
FD model for air conditioner (cooling).

The appliance has an automatic power management system that tries to keep the cooling capacity constant. This effect becomes evident when considering the decreasing dependence of I_1 on V_1. A similar effect can also be observed with a 17 W LED.

RMS [A] for vacuum cleaner (maximum power)	Phase [°] for vacuum cleaner (maximum power)
rms(I1) = 3E-3 V1 + 3.8	ph(I1) = 0
rms(I3) = 1.6 E-2 V1-2.83	ph(I3) = 0.45 Φ1 + 83
rms(I5) = 7.8E-3 V1-1.55	ph(I5) = −0.79 Φ1 + 342
rms(I7) = 5.5E-3 V1-1.12	ph(I7) = 1.38 Φ1-243
rms(I9) = 7E-4 V1-0.11	ph(I9) = 3.62 Φ1-894
rms(I11) = 8E-4 V1-0.15	ph(I11) = −3.17 Φ1 + 783
rms(I13) = 7E-05 V1 + 4.2E-2	ph(I13) = 1.32 Φ1-103
rms(I15) = 3E-4 V1-2.3E-2	ph(I15) = −0.14 Φ1-155
rms(I17) = 1E-3 V1-0.19	ph(I17) = −1.15 Φ1 + 395

Table 7.
FD model for a VACUUM CLEANER (maximum power).

RMS [A] for refrigerator (closed door)	Phase [°] for refrigerator (closed door)
rms(I1) = 2E-3 V1 + 0.48	ph(I1) = 0
rms(I3) = −5E-5 V1 + 1.2E-1	ph(I3) = 0.32 Φ1-196
rms(I5) = −4E-4 V1 + 1.3E-1	ph(I5) = 1.41 Φ1-388
rms(I7) = −4E-5 V1 + 1.9E-2	ph(I7) = 0.85 Φ1-274
rms(I9) = 5E-5 V1-6.1E-3	ph(I9) = −4.17 Φ1+ 840
rms(I11) = −3E-6 V1 + 2.7E-3	ph(I11) = 1.27 Φ1-229
rms(I13) = 1E-5 V1-8E-4	ph(I13) = −7.32 Φ1 + 1615
rms(I15) = 1E-5 V1-1.9E-3	ph(I15) = −3.95 Φ1 + 845
rms(I17) = −2E-5 V1 + 4.7E-3	ph(I17) = 2.95 Φ1-716

Table 8.
FD model for a refrigerator.

The FD models for other devices or other operating conditions are described in Ref. [10], also.

The content of the paragraph "2. Current sources models with parameters determined by measurements" reproduces the research reported for the first time in Ref. [10].

4. Frequency domain models for nonlinear devices with firing angle control devices

This section discusses the frequency domain models of a nonlinear load with a firing angle controller. The conduction state of devices such as thyristors, IGBTs (insulated gate bipolar transistors), TRIACs (triodes for alternating current), and DIACs (diodes for alternating current) is controlled by command signals [11].

4.1 Input-output models depending on firing angle and input voltage: Simulations

The models developed in the above sections characterize only the input behavior of a circuit block, which is a load in a power system. Sometimes, a load in the power system is a two-port: one port being its connection with the power system and the second port being associated with output quantities of interest. The models presented in this section can be developed on the basis of datasets obtained from network analyzer measurements or simulations with dedicated software.

In this section, we simulate a single-phase, fully controlled four-transistor bridge rectifier driven by a sinusoidal voltage source with amplitude V = 30 V and frequency f = 50 Hz (**Figure 12**). The load is a 10 ohms resistive load [12]. The following procedure can also be used for any type of nonlinear firing angle control device.

In this example, the firing angle is swept from 10 to 90 degrees with a 10-degree increment. The output port current (the load current) and input port current (the voltage source current) for all firing angles are shown in **Figures 13** and **14**.

The RMS value of the first five odd harmonic components of the input current varies with the firing angle α, as shown in **Figure 15**, and the RMS value of the next

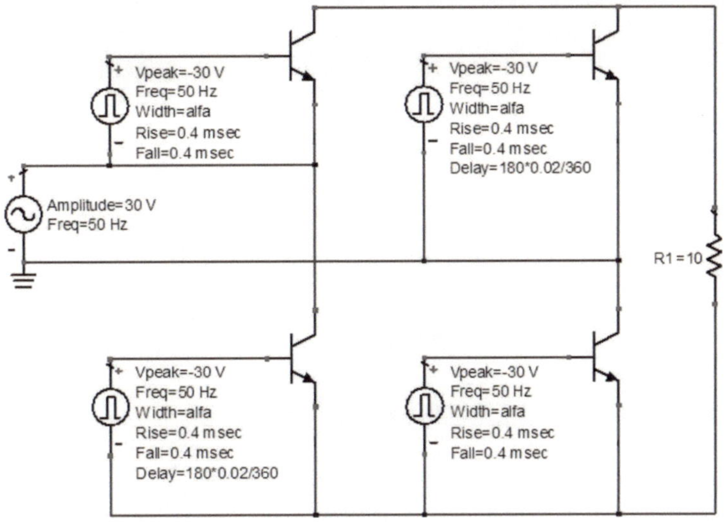

Figure 12.
Single-phase fully controlled bridge rectifier.

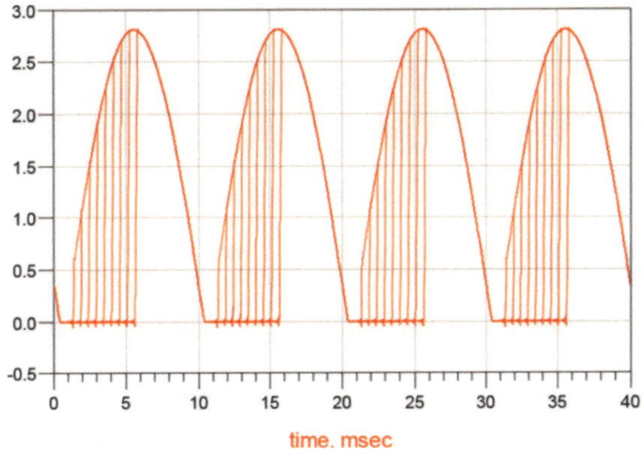

Figure 13.
The output port current.

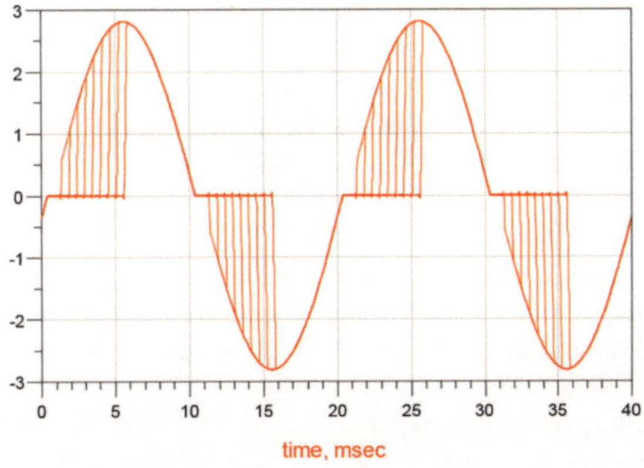

Figure 14.
The input port current.

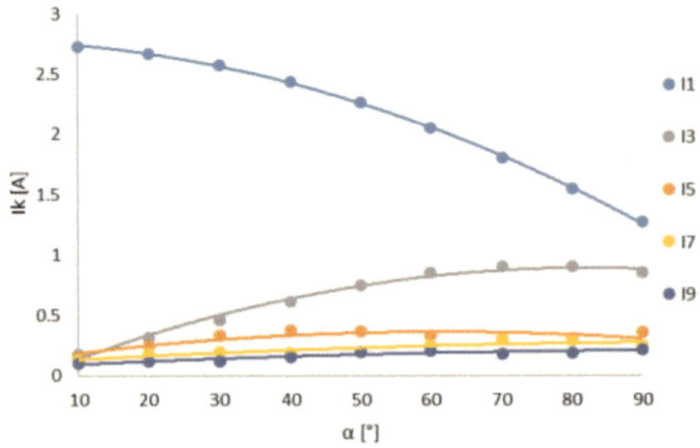

Figure 15.
RMS of odd current harmonics (1st-9th) vs. firing angle—Input port current.

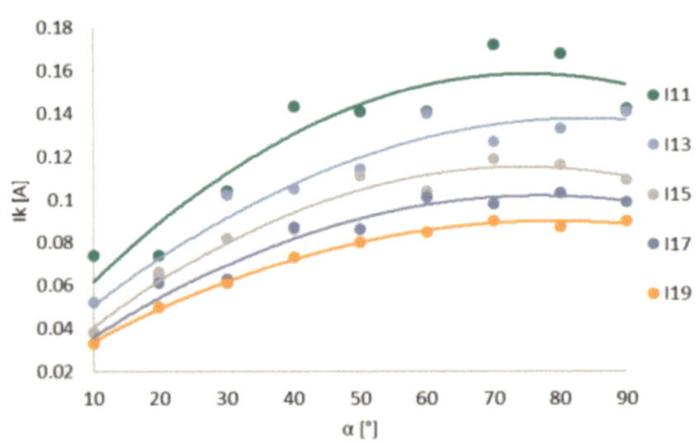

Figure 16.
RMS of odd current harmonics (11th–19th) vs. firing angle—Input port current.

five odd harmonic components of the input current varies with the firing angle α, as shown in **Figure 16**.

Starting from these samples, the α dependence for the most important odd input current harmonic RMS values has been computed by quadratic interpolation:

$\text{mag}(I_1) = -1.63\text{E-}4{\cdot}\alpha^2\text{-}2.25\text{E-}3{\cdot}\alpha + 2.78\text{E+}0.$

$\text{mag}(I_3) = -1.46\text{E-}4{\cdot}\alpha^2 + 2.39\text{E-}2{\cdot}\alpha - 8.59\text{E-}2.$

$\text{mag}(I_5) = -1.62\text{E-}5{\cdot}\alpha^2 + 3.38\text{E-}3{\cdot}\alpha + 9.87\text{E-}2.$

$\text{mag}(I_7) = -6.75\text{E-}5{\cdot}\alpha^2 + 8.22\text{E-}3{\cdot}\alpha + 1.09\text{E-}1.$

$\text{mag}(I_9) = -1.50\text{E-}5{\cdot}\alpha^2 + 2.91\text{E-}3{\cdot}\alpha + 6.37\text{E-}2.$

$\text{mag}(I_{11}) = -2.29\text{E-}5{\cdot}\alpha^2 + 3.43\text{E-}3{\cdot}\alpha + 2.95\text{E-}2.$

$\text{mag}(I_{13}) = -1.62\text{E-}5{\cdot}\alpha^2 + 2.70\text{E-}3{\cdot}\alpha + 2.50\text{E-}2.$

$\text{mag}(I_{15}) = -1.81\text{E-}5{\cdot}\alpha^2 + 2.68\text{E-}3{\cdot}\alpha + 1.54\text{E-}2.$

$\text{mag}(I_{17}) = -1.47\text{E-}5{\cdot}\alpha^2 + 2.26\text{E-}3{\cdot}\alpha + 1.47\text{E-}2.$

$\text{mag}(I_{19}) = -1.18\text{E-}5{\cdot}\alpha^2 + 1.86\text{E-}3{\cdot}\alpha + 1.64\text{E-}2.$

After a similar technique, the variation of the first five odd harmonic components of the phase input current with α is shown in **Figure 17**, and the variation of the next five odd harmonic components of the phase input current with α is shown in **Figure 18**.

From these samples, the phase dependence of the most important odd harmonics of the input current on α is calculated by quadratic interpolation:

$\text{phase}(I_1) = -2.35\text{E-}3{\cdot}\alpha^2\text{-}2.27\text{E-}1{\cdot}\alpha - 8.96\text{E+}1.$

$\text{phase}(I_3) = -5.88\text{E-}3{\cdot}\alpha^2\text{-}1.17\text{E+}0{\cdot}\alpha + 1.28\text{E+}2.$

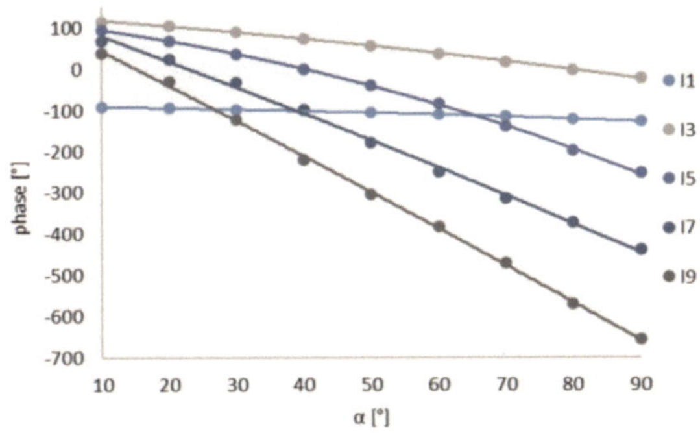

Figure 17.
Phase of odd current harmonics (1st-9th) vs. firing angle—Input port current.

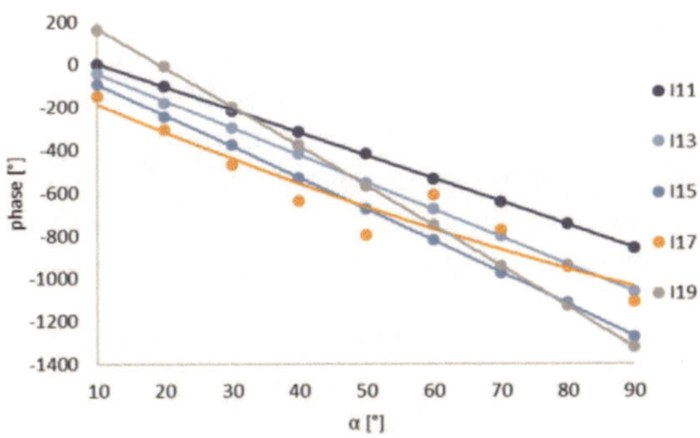

Figure 18.
Phase odd current harmonics (11th–19th) vs. firing angle—Input port current.

$$\text{phase}(I_5) = -2.52\text{E-}2\cdot\alpha^2\text{-}1.86\text{E+}0\cdot\alpha + 1.14\text{E+}2.$$
$$\text{phase}(I_7) = -6.10\text{E-}3\cdot\alpha^2\text{-}5.93\text{E+}0\cdot\alpha + 1.38\text{E+}2.$$
$$\text{phase}(I_9) = -5.44\text{E-}3\cdot\alpha^2\text{-}8.21\text{E+}0\cdot\alpha + 1.25\text{E+}2.$$
$$\text{phase}(I_{11}) = -6.83\text{E-}4\cdot\alpha^2\text{-}1.07\text{E+}1\cdot\alpha + 1.09\text{E+}2.$$
$$\text{phase}(I_{13}) = -1.36\text{E-}3\cdot\alpha^2\text{-}1.26\text{E+}1\cdot\alpha + 8.18\text{E+}1.$$
$$\text{phase}(I_{15}) = 3.35\text{E-}2\cdot\alpha^2\text{-}1.40\text{E+}1\cdot\alpha - 5.20\text{E+}1.$$
$$\text{phase}(I_{17}) = -7.93\text{E-}3\cdot\alpha^2\text{-}1.78\text{E+}1\cdot\alpha + 3.45\text{E+}2.$$
$$\text{phase}(I_{19}) = -4.11\text{E-}3\cdot\alpha^2\text{-}1.43\text{E+}1\cdot\alpha + 4.93\text{E+}1.$$

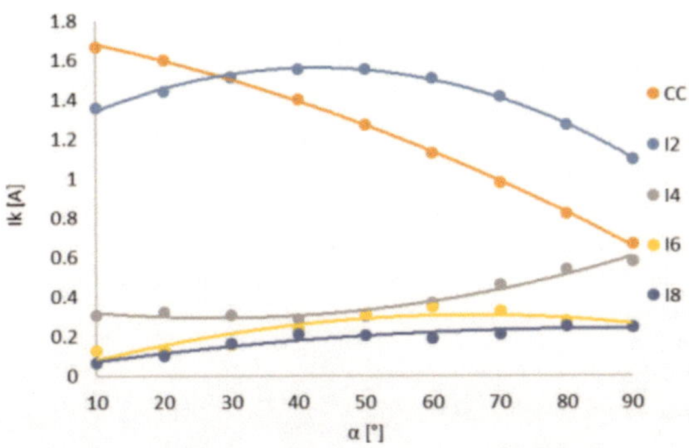

Figure 19.
RMS of even current harmonics (DC-8th) vs. firing angle—Output port current.

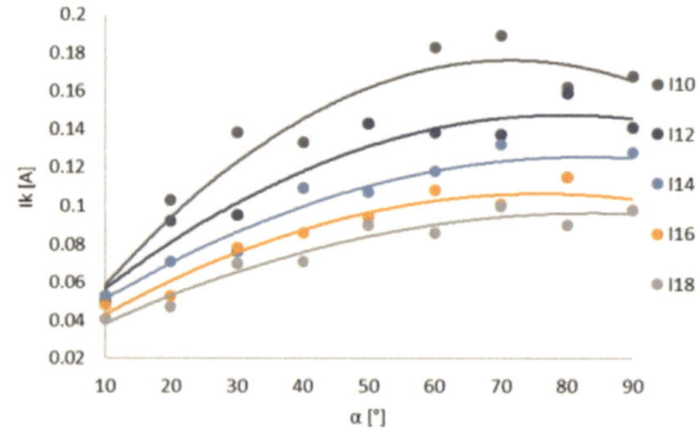

Figure 20.
RMS of even current harmonics (10th–18th) vs. firing angle—Output port current.

Similarly, the variation of the amplitude of the first five even harmonic components of the output current with the firing angle is shown in **Figure 19**, and the variation of the amplitude of the next five even harmonic components of the output current with the firing angle is shown in **Figure 20**.

The dependence of the most significant even output current harmonics amplitudes on the firing angle is calculated by quadratic interpolation based on the following samples:

$\mathrm{mag}(I_0) = -6.39\mathrm{E}\text{-}5\cdot\alpha^2\text{-}6.40\mathrm{E}\text{-}3\cdot\alpha + 1.75\mathrm{E}{+}0.$
$\mathrm{mag}(I_2) = -2.03\mathrm{E}\text{-}4\cdot\alpha^2 + 1.73\mathrm{E}\text{-}2\cdot\alpha + 1.19\mathrm{E}{+}0.$
$\mathrm{mag}(I_4) = 8.11\mathrm{E}\text{-}5\cdot\alpha^2\text{-}4.54\mathrm{E}\text{-}3\cdot\alpha + 3.55\mathrm{E}\text{-}1.$
$\mathrm{mag}(I_6) = -7.29\mathrm{E}\text{-}5\cdot\alpha^2 + 9.56\mathrm{E}\text{-}3\cdot\alpha - 8.38\mathrm{E}\text{-}3.$
$\mathrm{mag}(I_8) = -2.96\mathrm{E}\text{-}5\cdot\alpha^2 + 5.01\mathrm{E}\text{-}3\cdot\alpha + 2.43\mathrm{E}\text{-}2.$
$\mathrm{mag}(I_{10}) = -3.13\mathrm{E}\text{-}5\cdot\alpha^2 + 4.46\mathrm{E}\text{-}3\cdot\alpha + 1.69\mathrm{E}\text{-}2.$
$\mathrm{mag}(I_{12}) = -1.87\mathrm{E}\text{-}5\cdot\alpha^2 + 2.98\mathrm{E}\text{-}3\cdot\alpha + 2.88\mathrm{E}\text{-}2.$
$\mathrm{mag}(I_{14}) = -1.40\mathrm{E}\text{-}5\cdot\alpha^2 + 2.32\mathrm{E}\text{-}3\cdot\alpha + 2.93\mathrm{E}\text{-}2.$
$\mathrm{mag}(I_{16}) = -1.46\mathrm{E}\text{-}5\cdot\alpha^2 + 2.21\mathrm{E}\text{-}3\cdot\alpha + 2.22\mathrm{E}\text{-}2.$
$\mathrm{mag}(I_{18}) = -1.06\mathrm{E}\text{-}5\cdot\alpha^2 + 1.78\mathrm{E}\text{-}3\cdot\alpha + 2.16\mathrm{E}\text{-}2.$

The variation of the phase of the first five even harmonic components of the input current with the firing angle is shown in **Figure 21**, and the variation of the phase of the next five even harmonic components of the input current with the firing angle is shown in **Figure 22**.

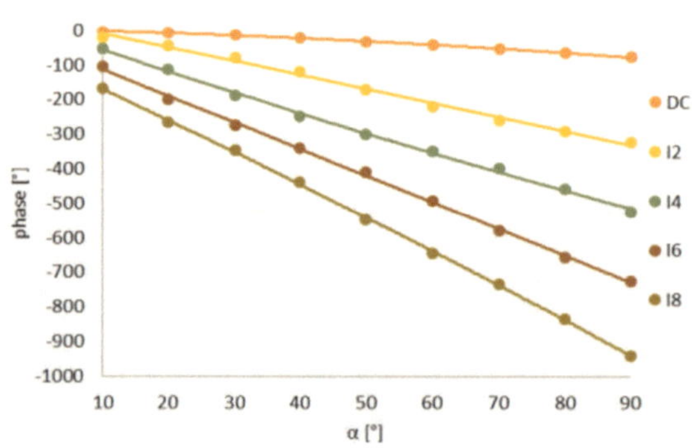

Figure 21.
Phase of even current harmonics (2nd-8th) vs. firing angle—Output port current.

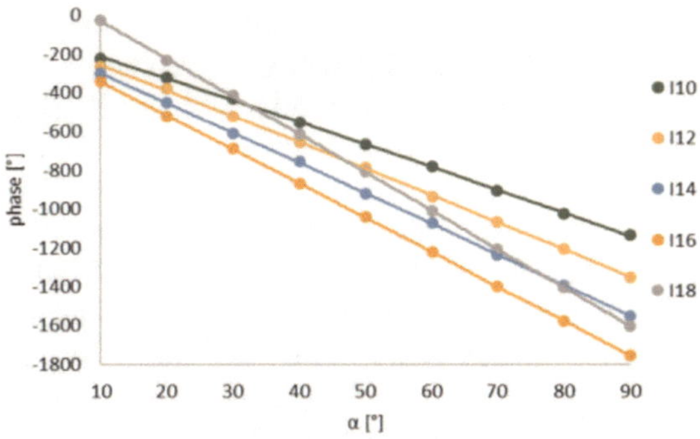

Figure 22.
Phase of even current harmonics (10th–18th) vs. firing angle—Output port current.

The phase dependence of the most important even harmonic output current components on α is calculated by quadratic interpolation starting from these data:

$phase(I_2) = -5.42E\text{-}3 \cdot \alpha^2 - 4.03E\text{-}1 \cdot \alpha + 2.64E{+}0.$

$phase(I_4) = -1.42E\text{-}3 \cdot \alpha^2 - 3.89E{+}0 \cdot \alpha + 2.99E{+}1.$

$phase(I_6) = 7.53E\text{-}3 \cdot \alpha^2 - 6.49E{+}0 \cdot \alpha + 8.47E{+}0.$

$phase(I_8) = -9.28E\text{-}4 \cdot \alpha^2 - 7.60E{+}0 \cdot \alpha - 3.69E{+}1.$

$phase(I_{10}) = -8.80E\text{-}3 \cdot \alpha^2 - 8.75E{+}0 \cdot \alpha - 8.11E{+}1.$

$phase(I_{12}) = -7.12E\text{-}3 \cdot \alpha^2 - 1.09E{+}1 \cdot \alpha - 1.05E{+}2.$

$phase(I_{14}) = -7.50E\text{-}3 \cdot \alpha^2 - 1.29E{+}1 \cdot \alpha - 1.27E{+}2.$

$phase(I_{16}) = -4.53E\text{-}3 \cdot \alpha^2 - 1.52E{+}1 \cdot \alpha - 1.46E{+}2.$

$phase(I_{18}) = -3.57E\text{-}3 \cdot \alpha^2 - 1.73E{+}1 \cdot \alpha - 1.70E{+}2.$

4.2 New input-output FD model implementation

Ideally, the grid provides a constant RMS voltage, such as 230 V. The voltage is considered a constant RMS value, even if it fluctuates within a small range. Therefore, we can say that, under normal conditions, circuits operate at constant magnitude values at and around the supply voltage, and the next model of the current source type is proposed. This model uses a current source controlled by the input voltage. Each current harmonic is described by its magnitude and phase. The control parameters are the amplitude and phase of the fundamental component of the input voltage and the firing angle. This is the first improvement of the model presented in this section. The second improvement of the same model is its implementation in ADS using a Frequency Domain Defined 2-Port component (FDD2P), and its validation by simulations leading to the same results as those obtained by the simulation of the circuit in **Figure 23**.

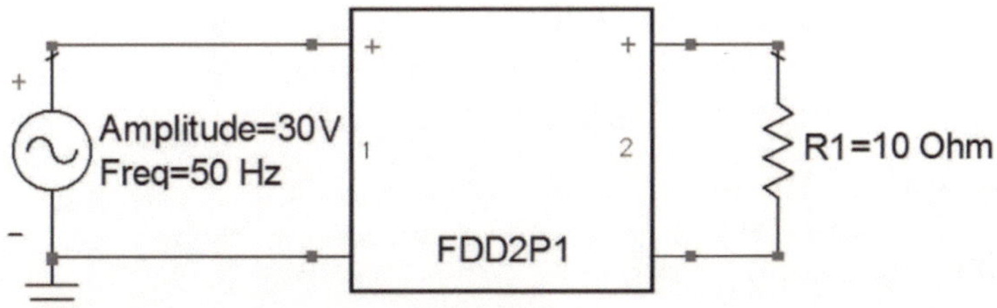

Figure 23.
Nonlinear controlled source model of circuit in ***Figure 12****.*

The equations obtained in the previous section and the magnitude and phase dependencies are implemented in the ADS software, as shown in **Figures 24–26**.

One major advantage of this type of FD model is the reduction of the simulation time. In **Table 9**, the CPU times for the simulations of the circuit in **Figure 12** using the TD model and using the FD model for the HB analysis are given.

An input-output FD model for a full-controlled bridge rectifier with four transistors has been presented in this section. The model consists of two sets of equations and can be extended to circuits with multiple ports. The models presented in this section can be obtained by using a network analyzer to make magnitude and phase measurements of all harmonic components of interest, or from datasets obtained using circuit simulation software that provides odd and even current harmonics. This model is characterized by a quadratic dependence of the amplitudes and phases of the current harmonics on the firing angle at the same input voltage. FD models reduce simulation time compared to classic HB analysis using TD models. To further improve the accuracy of the model, the number of considered harmonics can be increased, and the interpolation approximation function can be improved. However, the number of

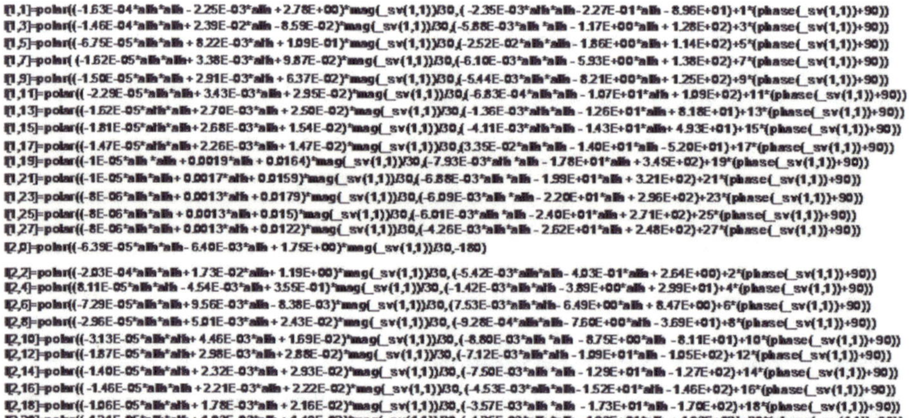

Figure 24.
Model implementation in ADS.

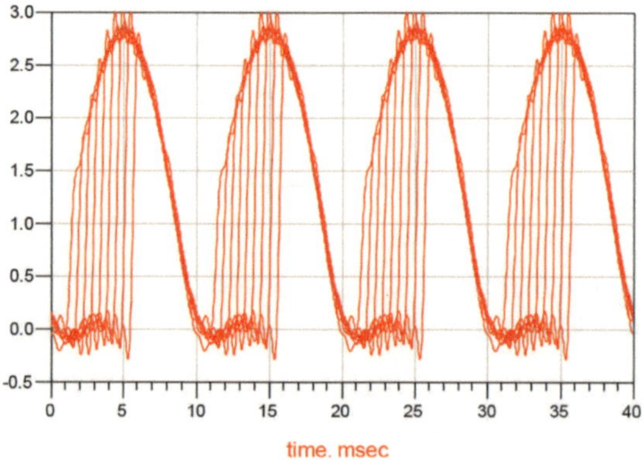

Figure 25.
The output port current—Simulated.

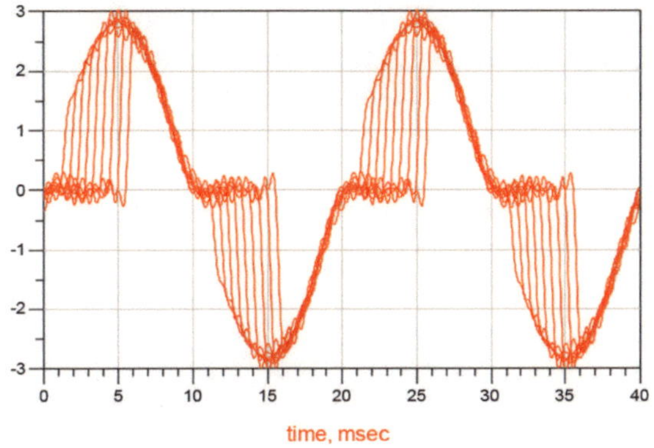

Figure 26.
The input port current—Simulated.

Model	CPU time [s]
TD	8.80
FD	2.37

Table 9.
Simulation results.

harmonics considered must be chosen carefully, as it increases simulation time without appreciably improving results. This type of FD model can also be used for larger circuits and networks. In this case, it is foreseeable that the simulation time will be significantly reduced compared to the classical model. Another important advantage is after the computation of the model parameters, a lot of new simulations can be performed in a considerably shorter time than a customary analysis.

4.3 Models depending on firing angle and input voltage: Measurements

Some measurements on one-phase and three-phase thyristor rectifiers (**Figures 27** and **28**), which may lead to their frequency domain models, are reported in this section together with an FD model of a one-phase rectifier with two thyristors.

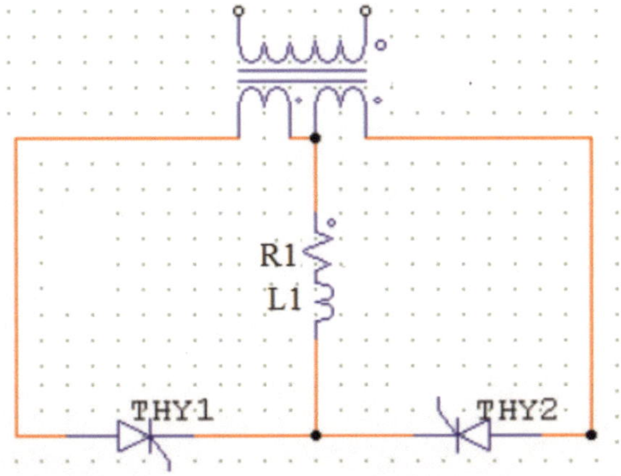

Figure 27.
One-phase rectifier.

The current harmonics amplitudes and phases, depending on the fundamental component of the voltage, are measured for various values of the firing angle (**Figures 29–38**) in the case of the one-phase rectifier.

The following measurement results are obtained for one phase of the three-phase rectifier in **Figures 28, 39–48**.

Similar results have been obtained for the other two phases of the rectifier.

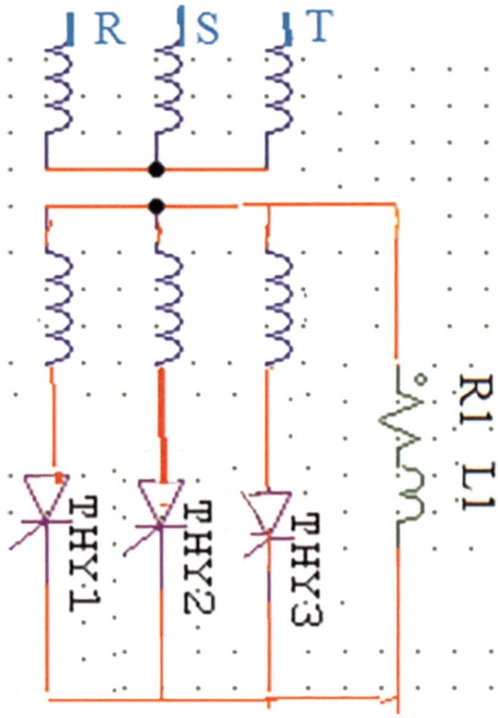

Figure 28.
Three-phase rectifier.

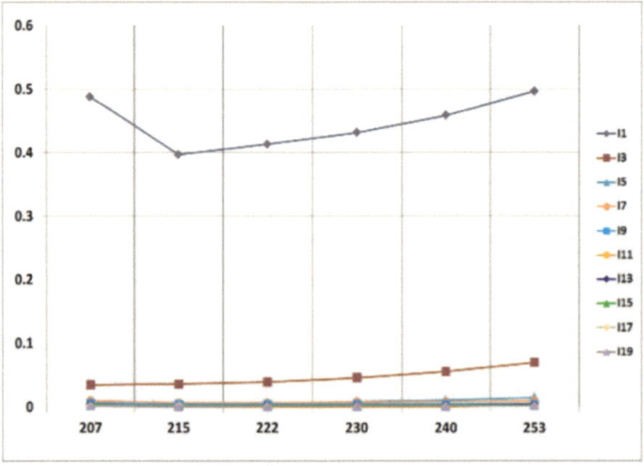

Figure 29.
RMS values for I1—I19 for α = 21°.

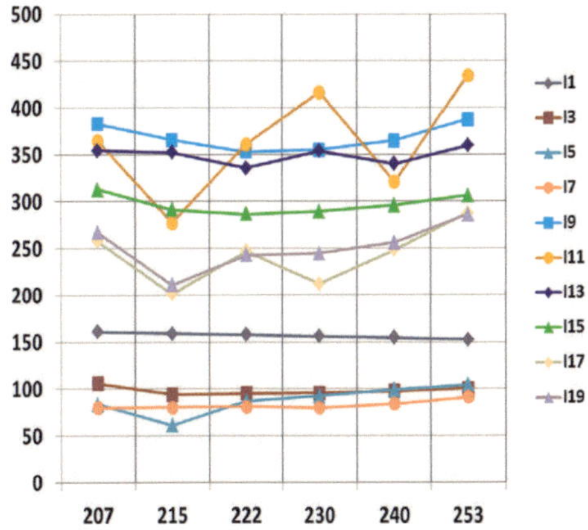

Figure 30.
PHASE values for I1—I19 α = 21°.

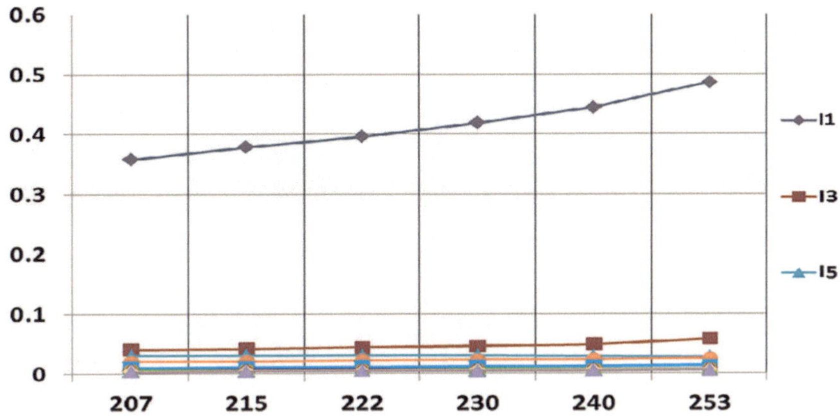

Figure 31.
RMS values for I1—I19 for α = 50°.

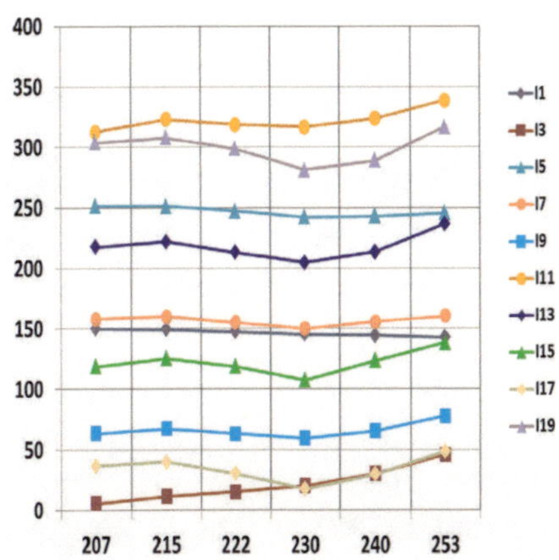

Figure 32.
PHASE values for I1—I19 for α = 50°.

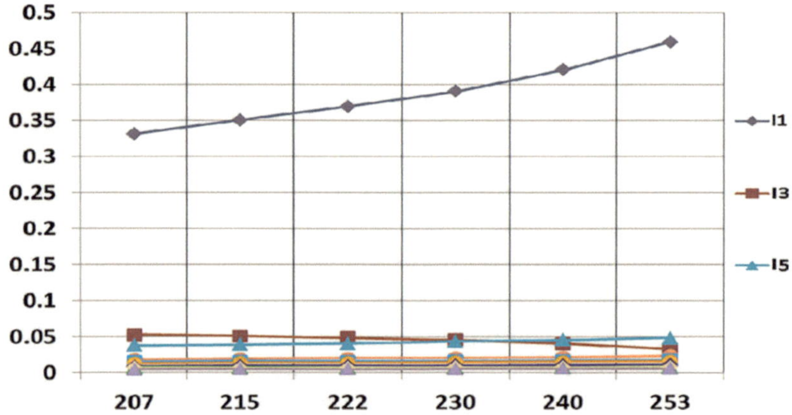

Figure 33.
RMS values for I1—I19 for α = 70°.

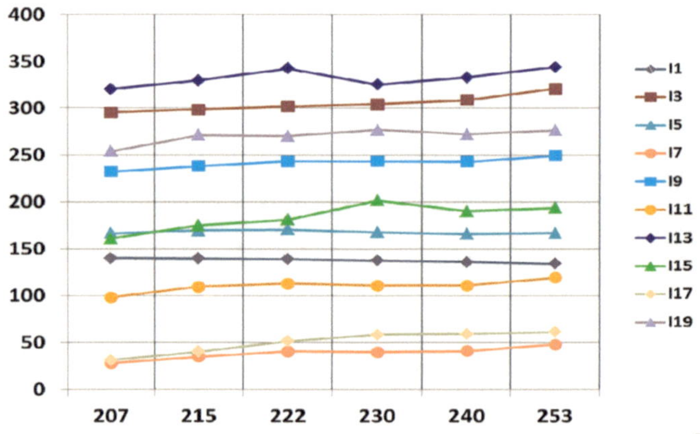

Figure 34.
PHASE values for I1—I19 for α = 70°.

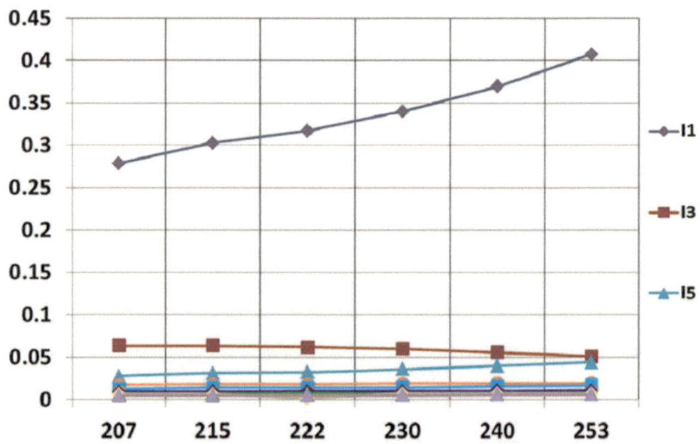

Figure 35.
RMS values for I1—I19 for α = 90°.

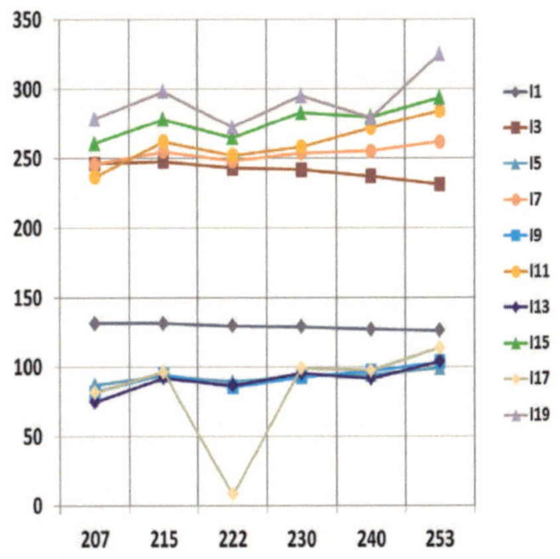

Figure 36.
PHASE values for I1—I19 for α = 90°.

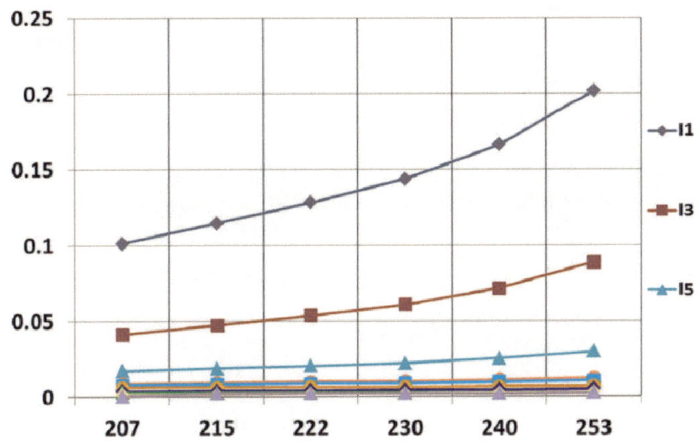

Figure 37.
RMS values for I1—I19 for α = 150°.

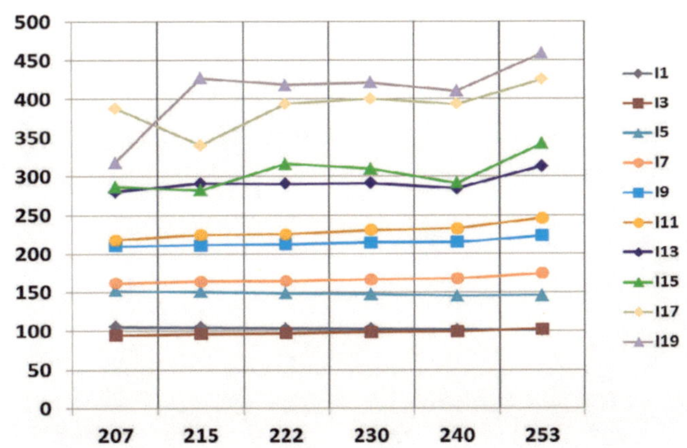

Figure 38.
PHASE values for I1—I19 for α = 150°.

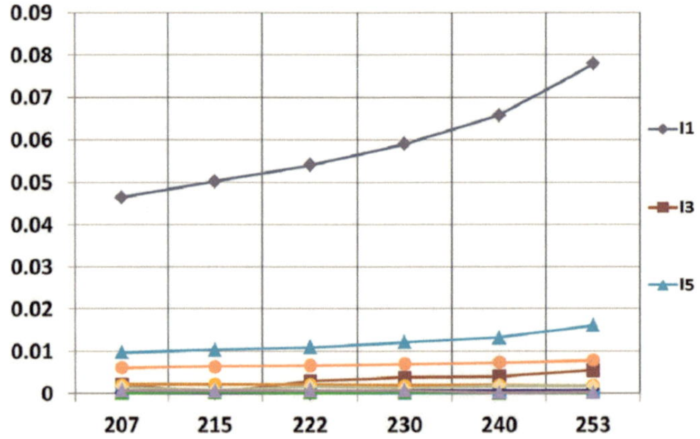

Figure 39.
RMS values for α = 21°.

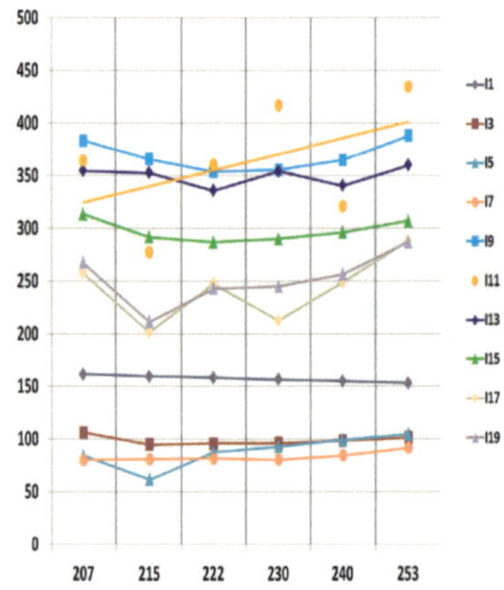

Figure 40.
PHASE values for α = 21°.

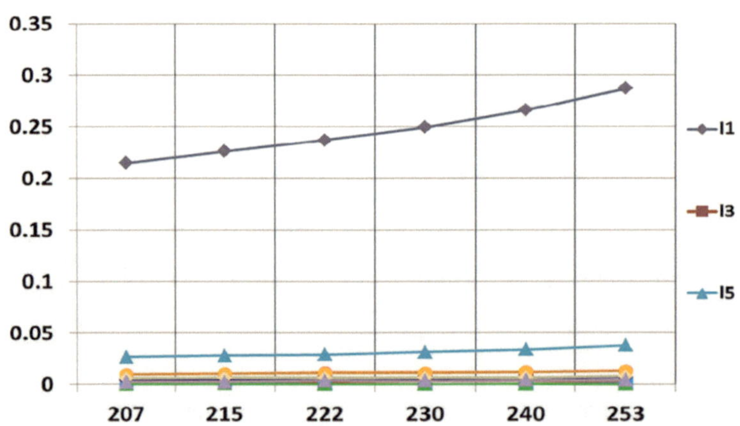

Figure 41.
RMS values for α = 50°.

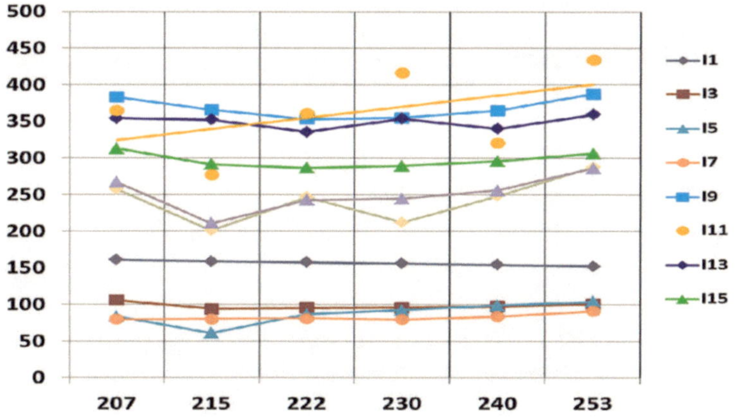

Figure 42.
PHASE values for α = 50°.

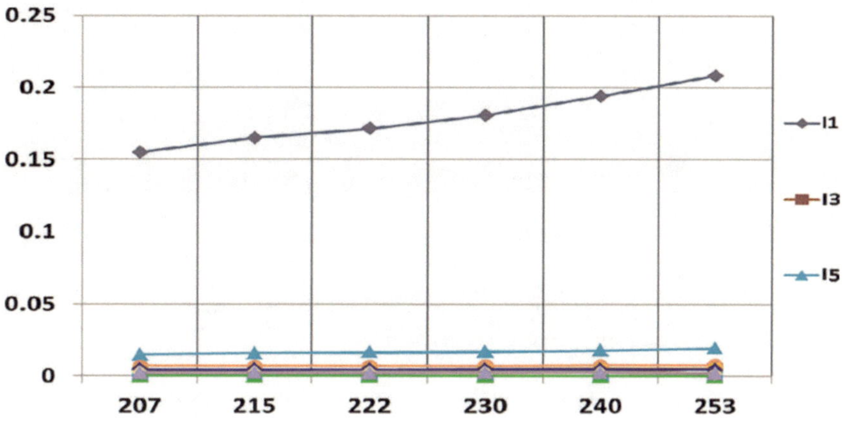

Figure 43.
RMS values for α = 70°.

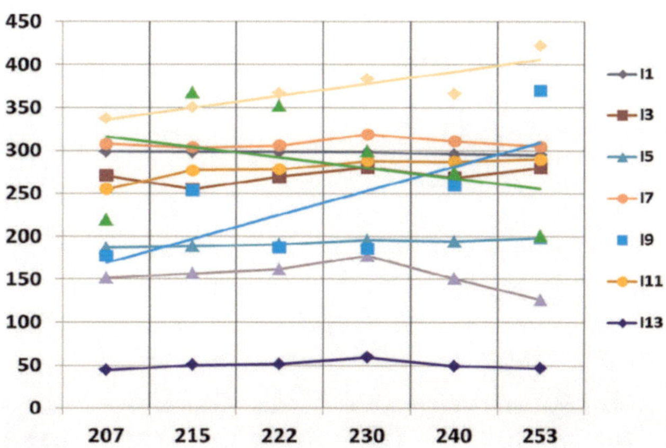

Figure 44.
PHASE values for α = 70°.

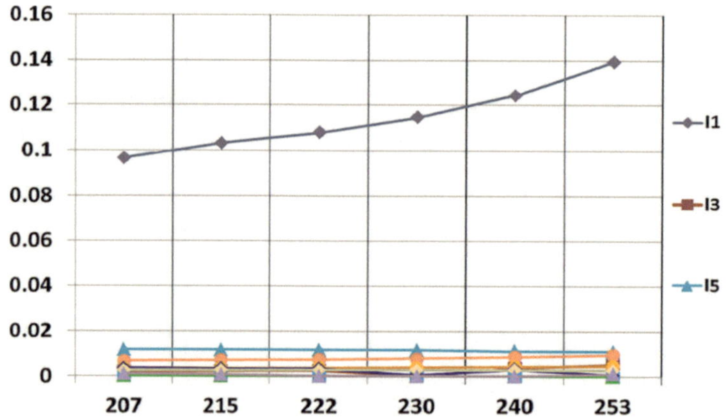

Figure 45.
RMS values for α = 90°.

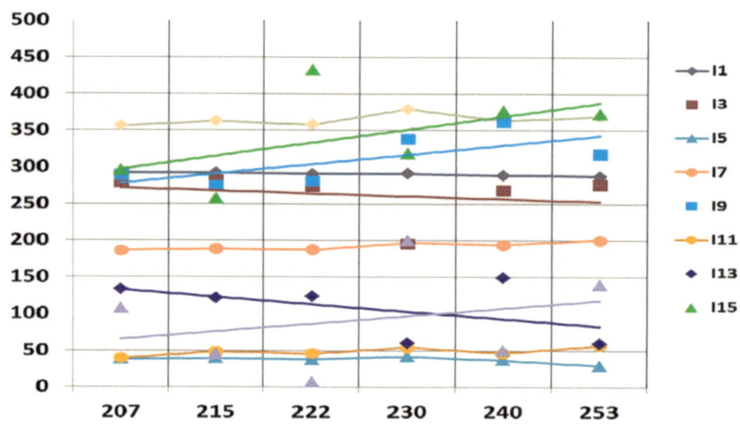

Figure 46.
PHASE values for α = 90°.

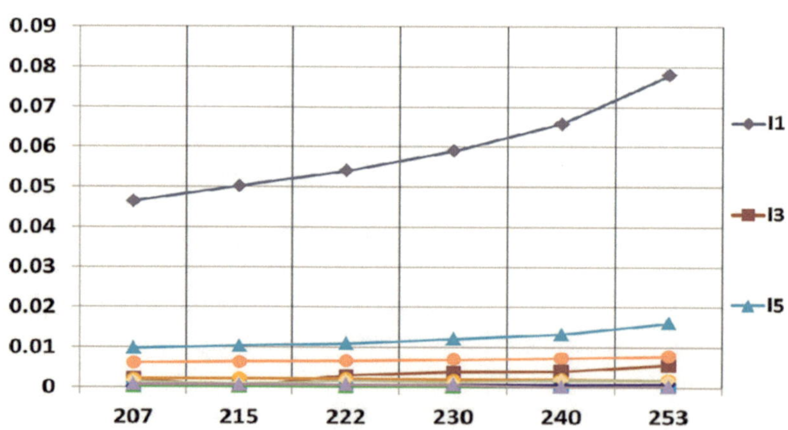

Figure 47.
RMS values for α = 121°.

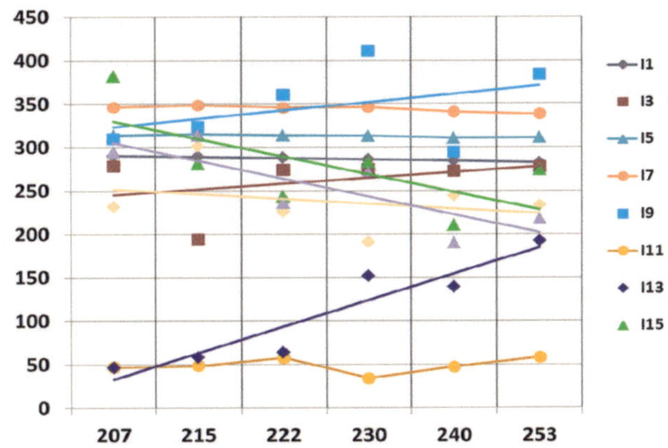

Figure 48.
PHASE values for α = 121°.

4.4 Frequency domain model of a one-phase two thyristors rectifier

From the measurement results, it follows that a useful model of a one-phase thyristor rectifier must compute the dependencies of the RMS and phase values of each current harmonic component on V_1 and *phase*(V_1), starting from the measured dependencies on V_1 for a set of firing angles. The simplest dependence on V_1 and *phase* (V_1) is a polynomial of two variables, which can be easily implemented in ADS software.

In order to verify the validity of our approach, we ignore the measured values for $\alpha_3 = 70^0$ and try to compute them using two interpolation polynomials. The current amplitudes and phases of the current harmonics have been measured for the following values of V_1 and α (**Tables 10** and **11**).

To verify the validity of our approach, we ignore the measured values for $\alpha_3 = 70^0$ and compute them using several interpolation methods:

- the poly45 function, which is a polynomial implemented in MATLAB and is based on the measured results for four values of α and five values of V_1,
- other interpolation polynomials, based on five values for α and six values of V_1.

U_1 [V]	U_2 [V]	U_3 [V]	U_4 [V]	U_5 [V]	U_6 [V]
207	215	222	230	240	253

Table 10.
V_1 values.

α1 [°]	α2 [°]	α3 [°]	α4 [°]	α5 [°]	α6 [°]
21	50	70	90	126	150

Table 11.
Firing angles values.

Finally, these results are compared with those given by a genetic algorithm based on the current samples measured by the network analyzer [13].

Interpolation algorithm: Giving the measured current samples $i_j = i(t_j), \mathrm{j} = \overline{1, S}$, the RMS values I_{2k-1} and initial phases φ_{2k-1} for the fundamental component and the odd harmonics up to the 11th order $(k = \overline{1, 6})$ are calculated so that the sum of the above components represents the best approximation of the measured current waveform:

$$i(t) = \sum_{k=1}^{6} \sqrt{2} I_{2k-1} sin\left((2k-1)\omega t + \varphi_{2k-1}\right) \tag{1}$$

To this end, the following error function is defined:

$$ERR(I_1, \varphi_1, I_3, \varphi_3, \dots, I_{11}, \varphi_{11}) = \sum_{i=1}^{S} \left[i_j - i(t_j)\right]^2 \tag{2}$$

Introducing (1) in (2) it follows:

$$ERR(I_1, \varphi_1, I_3, \varphi_3, \dots, I_{11}, \varphi_{11}) = \sum_{j=1}^{S} \left[i_j - \sum_{k=1}^{6} \sqrt{2} I_{2k-1} sin\left((2k-1)\omega t_j + \varphi_{2k-1}\right)\right]^2 \tag{3}$$

This error function has 12 variables, so a solution vector x with 12 components is defined to minimize the value of the above function using MATLAB. The corresponding relationship between the solution vector and each component of x is:

$$A_{2k-1} \to x_{2k-1}, \varphi_{2k-1} \to x_{2k}, k = \overline{1, 6} \tag{4}$$

Employing (4) in (3), we obtain:

$$ERR(x_1, x_2, \dots, x_{11}, x_{12}) = \sum_{j=1}^{S} \left[i_j - \sum_{k=1}^{6} \sqrt{2} x_{2k-1} sin\left((2k-1)\omega t_j + x_{2k}\right)\right]^2 \tag{5}$$

The error function in (5) was generated with MAPLE and converted into the MATLAB code. This error function value can be minimized using the MATLAB function ga from the Global Optimization Toolbox. This minimization method uses a genetic algorithm to find the RMS value and phase of the odd harmonic components of order 1–11. To do this, the ga function uses the following options: "HybridFcn", @fminunc, "Generations", 1200, "TolFun", 1e-15.

In this way, the algorithm calculates the RMS and initial phase values of the six odd harmonic current components for a set of $M \times N$ points defined by the values of the fundamental voltage components $U_m, m = \overline{1, M}$ and the values of firing angle $\alpha_n, n = \overline{1, N}$.

To calculate the same RMS and initial phase values of the six odd harmonic components at the new point, the polynomial interpolation described below is used. That is, in order to calculate the RMS value I_k and the initial phase φ_k of the kth current harmonic, two $(M-1) \times (N-1)$-order interpolation polynomials are built. Here are the two variable polynomials in U and α:

$$\begin{aligned} P_{I_k}(U,\alpha) &= \sum_{i=1}^{M}\left(U^{i-1}\sum_{j=1}^{N}A_{j+(i-1)\cdot N}\alpha^{j-1}\right) \\ &= A_1 + A_2\alpha + \dots A_N\alpha^{N-1} + A_{N+1}U + \dots + A_{2N}U\alpha^{N-1} + \dots \\ &\quad + A_{1+(M-1)\cdot N}U^{M-1} + \dots + A_{M\cdot N}U^{M-1}\alpha^{N-1} \end{aligned} \tag{6}$$

$$\begin{aligned} P_{\varphi_k}(U,\alpha) &= \sum_{i=1}^{M}\left(U^{i-1}\sum_{j=1}^{N}B_{j+(i-1)\cdot N}\alpha^{j-1}\right) \\ &= B_1 + B_2\alpha + \dots B_N\alpha^{N-1} + B_{N+1}U + \dots + B_{2N}U\alpha^{N-1} + \dots \\ &\quad + B_{1+(M-1)\cdot N}U^{M-1} + \dots + B_{M\cdot N}U^{M-1}\alpha^{N-1} \end{aligned} \tag{7}$$

These polynomials are constructed in such a way that their computation yields the RMS value of I_k and the initial phase of I_k, which correspond to the fundamental voltage component U_m and firing angle α_n:

$$P_{I_k}(U_m,\alpha_n) = \sum_{i=1}^{M}\left(U_m^{i-1}\sum_{j=1}^{N}A_{j+(i-1)\cdot N}\alpha_n^{j-1}\right) = I_k(U_m,\alpha_n) \tag{8}$$

$$P_{\varphi_k}(U_m,\alpha_n) = \sum_{i=1}^{M}\left(U_m^{i-1}\sum_{j=1}^{N}B_{j+(i-1)\cdot N}\alpha_n^{j-1}\right) = \varphi_k(U_m,\alpha_n) \tag{9}$$

The computation of the coefficients $A_{j+(i-1)\cdot N}$, respectively $B_{j+(i-1)\cdot N}$, $i = \overline{1,M}$, $j = \overline{1,N}$ is done by solving the following equation systems:

$$\begin{aligned} &A_1 + A_2\alpha_n + \dots + A_{j+(i-1)\cdot N}U_m^{i-1}\alpha_n^{j-1} + \dots + A_{M\cdot N}U_m^{M-1}\alpha_n^{N-1} = I_k(U_m,\alpha_n), \\ &m = \overline{1,M}, n = \overline{1,N}, \\ &\left(i = \overline{1,M}, j = \overline{1,N}\right) \end{aligned} \tag{10}$$

$$\begin{aligned} &B_1 + B_2\alpha_n + \dots + B_{j+(i-1)\cdot N}U_m^{i-1}\alpha_n^{j-1} + \dots + B_{M\cdot N}U_m^{M-1}\alpha_n^{N-1} = \varphi_k(U_m,\alpha_n), \\ &m = \overline{1,M}, n = \overline{1,N}, \\ &\left(i = \overline{1,M}, j = \overline{1,N}\right) \end{aligned} \tag{11}$$

Knowing the values of the coefficients $A_{j+(i-1)\cdot N}$, and $B_{j+(i-1)\cdot N}$, for $i = \overline{1,M}$, and $j = \overline{1,N}$, the algorithm computes the RMS value and the initial phase of the kth current harmonic in a point corresponding to the fundamental voltage *RMS* value U', $U_1 \le U' \le U_M$, and to the firing angle α', $\alpha_1 \le \alpha' \le \alpha_N$, as:

$$I_k(U',\alpha') = P_{I_k}(U',\alpha') = \sum_{i=1}^{M}\sum_{j=1}^{N}A_{j+(i-1)\cdot N}\alpha'^{j-1}U'^{i-1} \tag{12}$$

$$\varphi_k(U',\alpha') = P_{\varphi_k}(U',\alpha') = \sum_{i=1}^{M}\sum_{j=1}^{N}B_{j+(i-1)\cdot N}\alpha'^{j-1}U'^{i-1} \tag{13}$$

In order to compare the results, the magnitudes in (12) and (13) are computed with the function fit from the Curve Fitting Toolbox of MATLAB. The parameter fitType is "poly45", and in this case, the polynomial has the order 4×5, its variables being α_n U_m. Obviously, the FD models are the curves $I_k(U_m, \alpha_n)$ $\alpha_n = ct$ and $\varphi_k(U_m, \alpha_n)$ $\alpha_n = ct$ obtained from (8) and (9) or from (12) and (13). These are polynomials in U_m and may be considered as a generalization of the linear models in (1). This kind of model can be easily implemented in ADS.

Results: The performance of the model obtained using the Poly45 function (MATLAB) and the polynomial interpolation described in the previous section is shown below. The RMS values of the dependence of the fundamental current components on U and α are represented by the red dots in **Figure 49**.

This dependence is compared to that of the genetic algorithm in Ref. [14], based on the measured time samples (green dots). If there is no measurement data for $\alpha = 70^0$, the measurement points correspond to 6 values of U and 5 values of α, so the poly45 function chooses the best 20 points to create the interpolating polynomial (blue points).

The error measure for this interpolation method is the distance between each pair of green and red points in **Figure 49**.

Figure 50 shows a similar interpolation error measurement for the initial phase of the fundamental component of the current computed using poly45 (MATLAB).

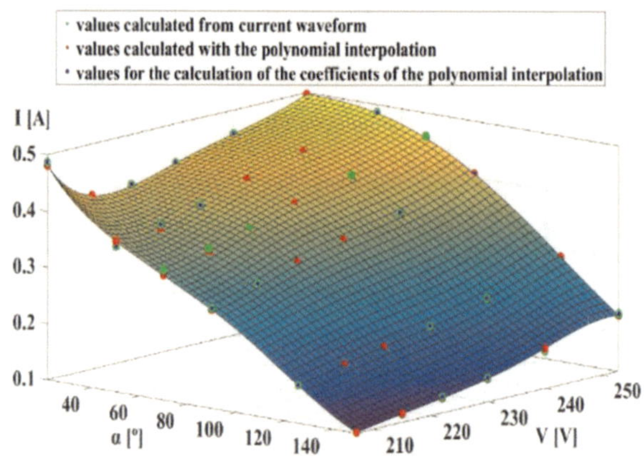

Figure 49.
First harmonic RMS value vs. U and α (poly45).

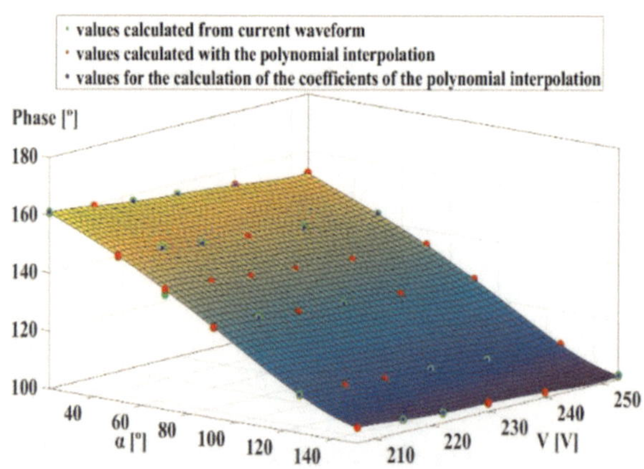

Figure 50.
First harmonic initial phase value vs. U and α (poly45).

The error of the proposed interpolation method can be estimated by analyzing **Figures 51** and **52**. A simple inspection of the distance between each pair of green and red points shows that the proposed interpolation method gives better results than the poly45 function (MATLAB).

Below, the RMS values and the initial phase values dependences on *U* of all odd current harmonics obtained by the two interpolation algorithms with $\alpha = 70^0$ are compared with the same dependences given by the genetic algorithm in Ref. [14], which are based on the measured time samples and is regarded as the minimum error data (**Figures 53–58**).

A simple inspection of **Figures 49–52** shows that the results obtained using the poly45 function (MATLAB) and the proposed interpolation algorithm are very close to those of the genetic algorithm based on measurement time samples in Ref. [14]. From **Figures 56–58**, it can be seen that the initial phase obtained using the poly45 function (MATLAB) and the proposed interpolation has a significant error compared to the initial phase obtained by the genetic algorithm in Ref. [14], unlike the RMS values.

Another possibility for different methods of error estimation is waveform reconstruction. **Figure 59** shows the following current waveforms: measured (red line), reconstructed using the algorithm in Ref. [14] (blue line), reconstructed using Poly45

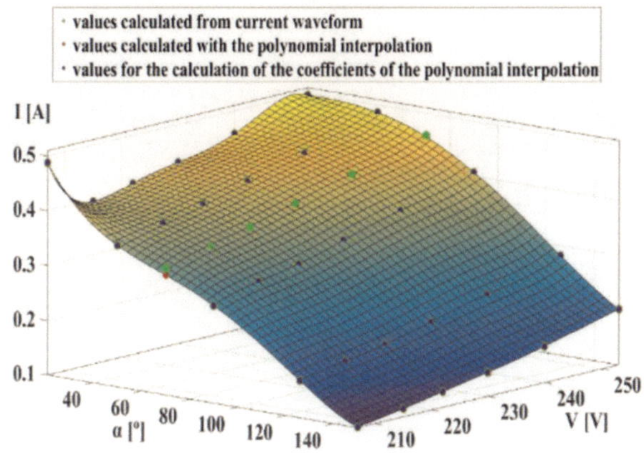

Figure 51.
First harmonic RMS value vs. U and α (proposed interpolation).

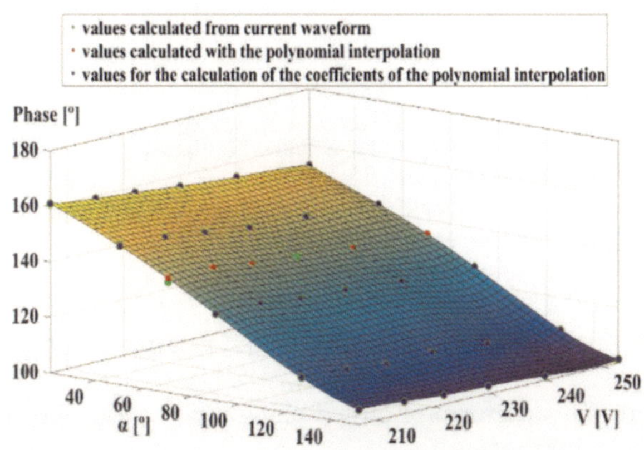

Figure 52.
First harmonic initial phase value vs. U and α (proposed interpolation).

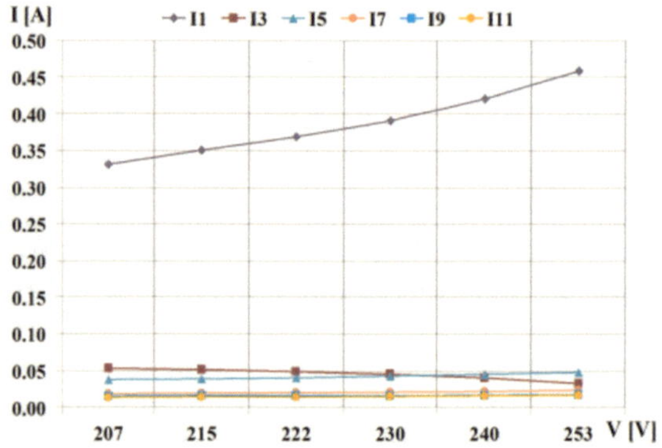

Figure 53.
RMS values vs. U for α = 70° obtained with the genetic algorithm [14].

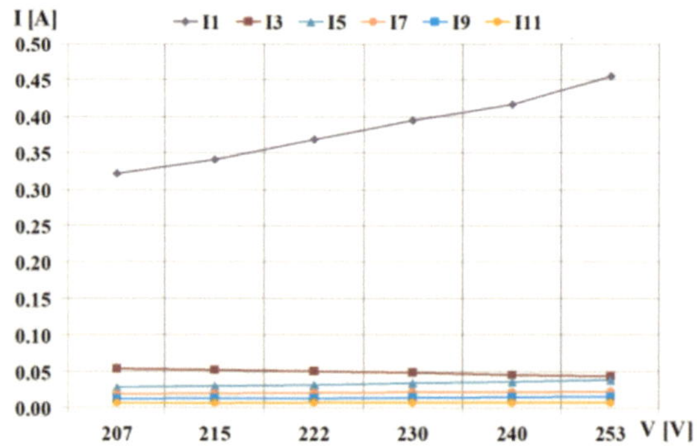

Figure 54.
RMS values vs. U for α = 70° obtained with poly45 (MATLAB).

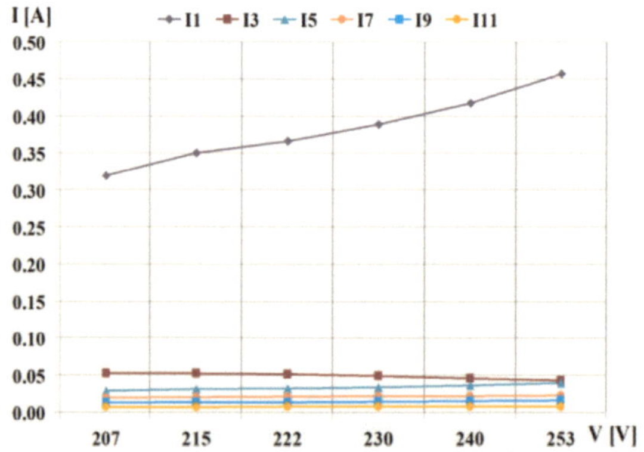

Figure 55.
RMS values vs. U for α = 70° obtained with the proposed interpolation.

interpolation (pink line), and reconstructed using the proposed polynomial interpolation (green line).

It is obvious that the waveform obtained using the genetic algorithm in Ref. [14] is closest to the measured waveform.

To measure such errors, **Figure 60** shows the spectrum of RMS values at $\alpha = 70^0$ and $U = 230$ V obtained using the poly45 interpolation algorithm (red line) and the

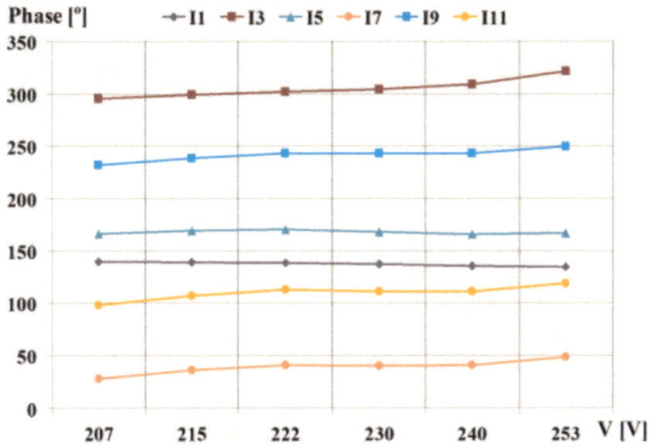

Figure 56.
Phases vs. U for α = 70° obtained with the genetic algoritm [13].

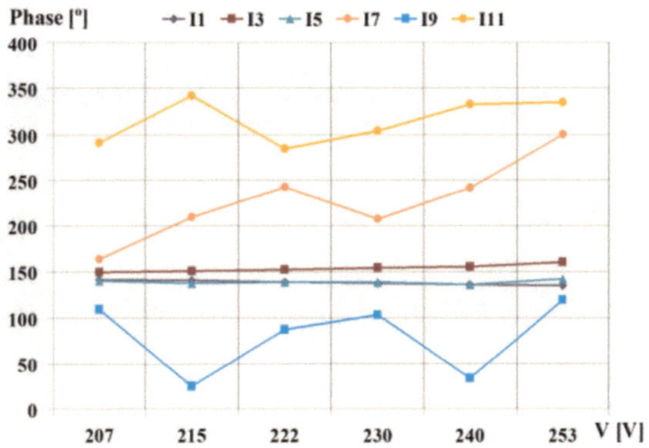

Figure 57.
Phases vs. U for α = 70° obtained with poly45 (MATLAB).

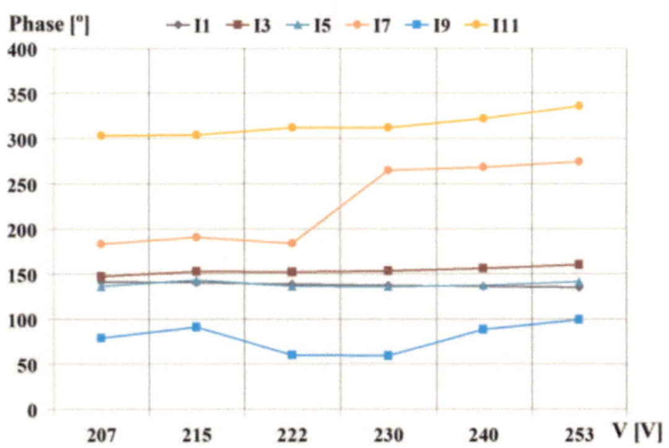

Figure 58.
Phases vs. U for α = 70° obtained with the proposed interpolation.

proposed interpolation algorithm (green line) in comparison with the spectrum (blue line) corresponding to the RMS value calculated by the algorithm in Ref. [14]. At first glance, all results are very close.

The phase spectrum at $\alpha = 70^0$ and $U = 230$ V calculated using the same algorithm is shown in **Figure 61**. In this case, some important errors can be noted.

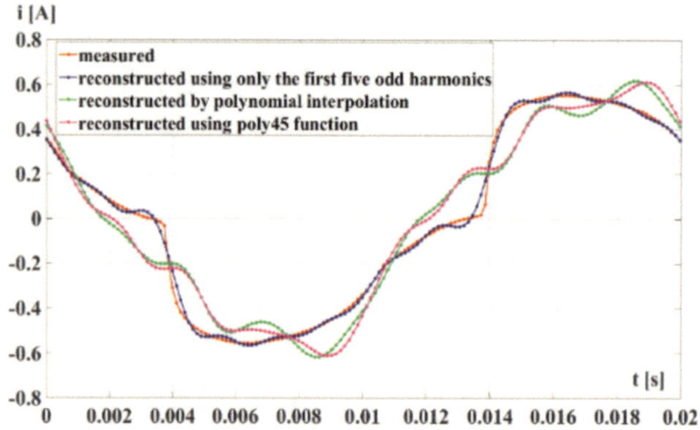

Figure 59.
Measured and reconstructed current waveforms for α = 70° and U = 230 V.

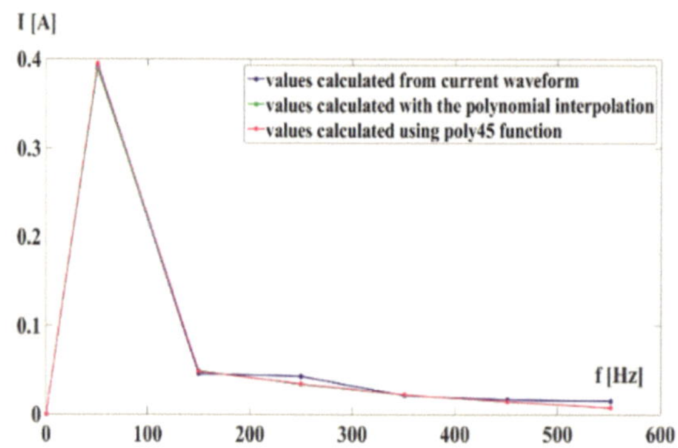

Figure 60.
RMS current harmonics spectrum.

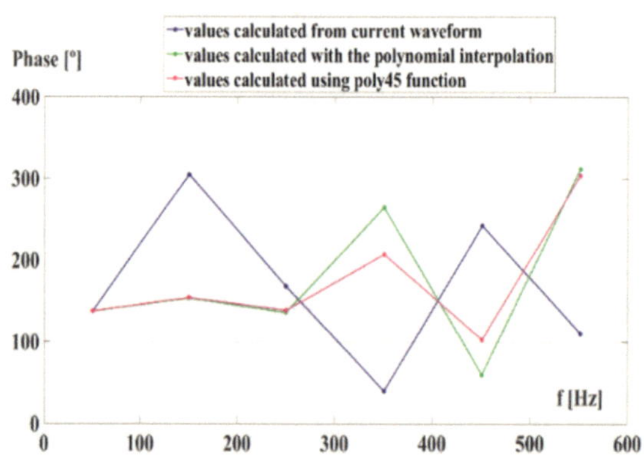

Figure 61.
Initial phase current harmonics spectrum.

4.5 Comparison between interpolation algorithms

Since the losses in the transmission line are proportional to I^2 (I – RMS value of the current), this value can be used as a failure criterion for the algorithm to calculate the current harmonics. The RMS current value can be calculated using the well-known formula:

	Genetic algorithm [8]	Proposed interpolation algorithm	"poly45" (MATLAB)
I_1 [mA]	390.471	389.063	394.656
I [mA]	396.683	394.537	400.050

Table 12.
RMS values computed from I_k.

	Measured	Genetic algorithm [8]	Proposed interpolation algorithm	Interpolation using "poly45"
I[mA]	396.925	396.395	394.676	400.285

Table 13.
RMS values computed from time samples.

$$I = \sqrt{\sum\nolimits_{k=1}^{6} I_{2k-1}^2} = \sqrt{I_1^2 + I_3^2 + I_5^2 + I_7^2 + I_9^2 + I_{11}^2} \tag{14}$$

The values of I_1 and I obtained using various algorithms using (14) are given in **Table 12.**

The RMS value of I can be calculated using time samples of the current waveform (**Table 13**).

$$i_j = i(t_j), j = \overline{1, S} \tag{15}$$

$$\text{as} : I = \sqrt{\frac{1}{S}\sum\nolimits_{j=1}^{S} i_j^2} \tag{16}$$

The maximum relative error between the RMS values of I calculated with different algorithms is 1.5%, so these results can be considered similar from a practical point of view.

Unlike diode rectifiers, whose behavior is determined only by the parameters V_1 and $phase(V_1)$, the behavior of a rectifier with two thyristors is determined by an additional parameter: the firing angle α. For certain values of α (in the case of the two thyristors rectifier for α = 21^0), the dependence of I_1 on U is nonlinear, so the linear small-signal model used in Refs. [8–10] is not valid. This is because the building of the FD model requires a relatively complex interpolation algorithm, as described in this section. The model consists of higher-order polynomials representing the surface, as shown in **Figures 49–52**, cross sections α = ct. Further research will be devoted to the FD models of the IGBT circuits.

Since the FD model of the diode rectifier achieves an order of magnitude reduction in CPU time compared to the HB analysis using the TD diode model [8], it is expected that the FD model of the firing angle control devices will lead to a similar reduction in simulation time.

The content of paragraph "3. Frequency domain models for nonlinear devices with firing angle control devices" reproduces the research reported for the first time in Refs. [12, 15].

5. FD and TD models efficiency

A three-phase test circuit with 10 identical loads (**Figure 62**), each consisting in 10 rectifiers, connected to a three-phase sinusoidal source with f = 50 Hz and amplitude

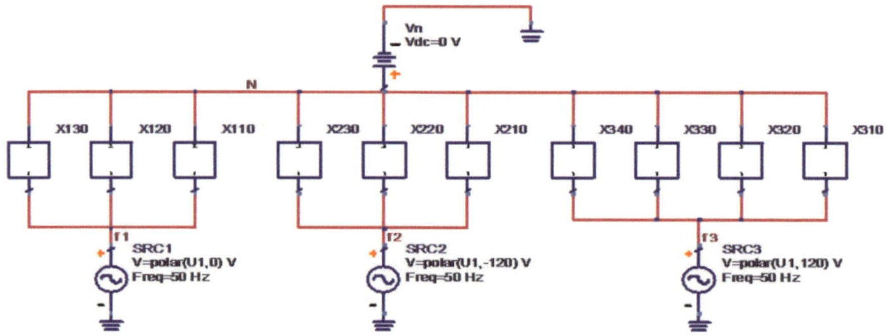

Figure 62.
Three-phase test circuit.

V = 230 V has been used for a comparison of the simulation times obtained with HB method of ADS using the FD models proposed in Ref. [8], HB method of ADS using the usual TD models, and the following analyses implemented in CADENCE using TD models: TRAN, PSS TRAN, and PSS-HB [16].

Each rectifier is connected to the 400 m line by a 6mm^2 cable of 30 m with R_{phase} = 4.767 Ω/Km and X_{phase} = 0.101 Ω/Km. Each sinusoidal source is connected to the load by a line of 400 m with R_{line} = 0.045 Ω/Km and X_{line} = 0.074 Ω/Km. Each group of loads contains five one-diode rectifiers and five two-diode rectifiers, as those in **Figures 1** and **4**. This home appliance is a non-equilibrated circuit containing 150 capacitors, 150 diodes, 100 inductors, and 200 resistors (**Figure 62**).

The phase currents $i_1(t)$, $i_2(t)$, $i_3(t)$ are pointed out in **Figures 63** and **64**, and $i_N(t)$ is considered through V_n.

The models used to obtain the results in **Figures 63–65** are based on linear dependence outlined in Section 1. To ensure the convergence of all analyses, we added the conductance Gmin = 1e-6 Ω^{-1} in parallel with all P-N junctions. The CPU times for all analyses are given in **Table 14** [16].

From **Table 14** it is obvious that the HB ADS analysis using the proposed models is the most efficient. Using FD models, the results corresponding to the original circuit (Gmin = 0) can be obtained in the same CPU time as those in **Table 14**. To obtain the same results, a certain number of iterations using decreasing values for Gmin must be performed employing Cadence analyses [17]. This certifies that, at least for this

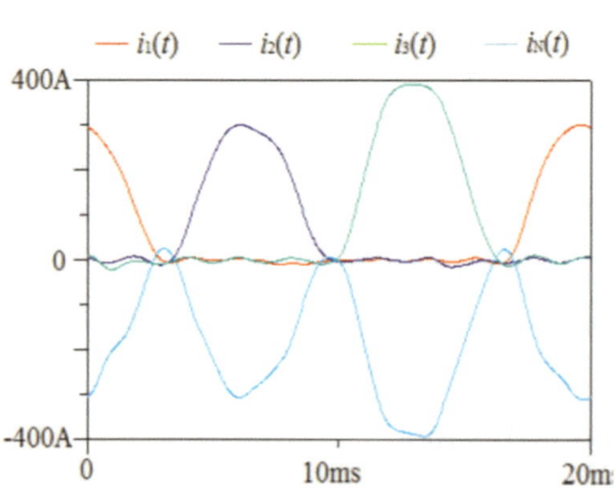

Figure 63.
Currents with HB (ADS) FD models [16].

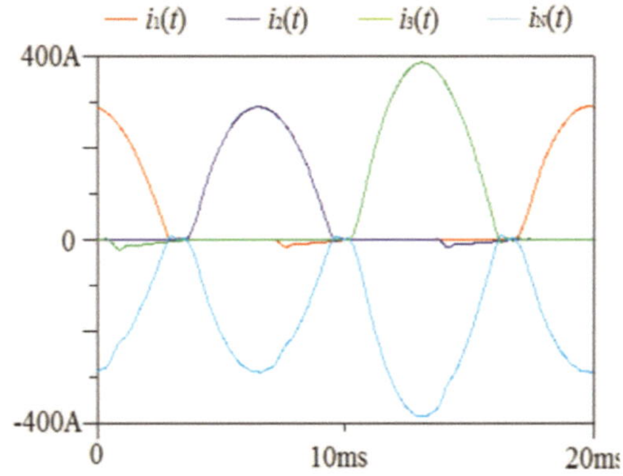

Figure 64.
Currents with HB (ADS) FD models [16].

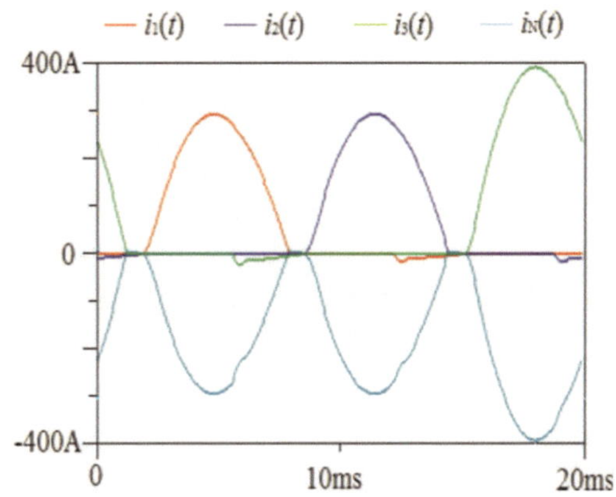

Figure 65.
Currents with PSS-HB (Cadence) TD models [16].

Analysis	CPU time [s]	Peak memory used [Mbytes]
HB ADS – FD MODELS	1.36	not available
HB ADS – TD MODELS	550.34	not available
Tran ADS – TD MODELS	10.76	not available
PSS-HB Cadence – TD MODELS	215	505
PSS tran Cadence – TD MODELS	5.39	241
Tran Cadence – TD MODELS	6.88	59.6

Table 14.
Simulation results.

example, the HB analysis with ADS using our proposed models is the most efficient. Other examples must be analyzed to prove if this statement is generally valid.

The content of paragraph "4. FD and TD models efficiency" reproduces the research reported for the first time in Ref. [16].

Author details

Florin Constantinescu[1], Alexandru Gabriel Gheorghe[1*], Mihai Eugen Marin[1] and Florin Roman Enache[2]

1 Polytechnic University of Bucharest, Bucharest, Romania

2 Military Technical Academy "Ferdinand I", Bucharest, Romania

*Address all correspondence to: alexandru.gheorghe@upb.ro

References

[1] IEEE recommended practices and requirements for harmonic control in electrical power systems. In: IEEE Std 519-1992. 9 April 1993. pp. 1-112. DOI: 10.1109/IEEESTD.1993.114370. Available from: https://ieeexplore.ieee.org/

[2] Pandi VR, Zeineldin HH, Xiao W. Determining optimal location and size of distributed generation resources considering harmonic and protection coordination limits. IEEE Transactions on Power Systems. 2013;**28**(2):1245-1254

[3] Hammad AE, Kamwa I, Viarouge P, Le-Huy H, Dickinson EJ, Medina A, et al. Modeling and simulation of the propagation of harmonics in electric power networks. I: Concepts, models, and simulation techniques. IEEE Transactions on Power Delivery. 1996; **11**:452-465

[4] Kundert KS, Sangiovanni-Vincentelli A. Simulation of nonlinear circuits in the frequency domain. IEEE Transactions on Computer-Aided Design. 1986;**CAD-5**(4): 521-535

[5] Medina A, Segundo-Ramirez J, Ribeiro P, Xu W, Lian KL, Chang GW, et al. Harmonic analysis in frequency and time domain. IEEE Transactions on Power Delivery. 2013;**28**(3):1813-1821

[6] Nassif A. Modeling, measurement and mitigation of power system harmonics [PhD thesis] in electrical and computer engineering, University of Alberta, Edmonton, Canada. 2009

[7] Yong J, Chen L, Nassif A, Xu W. A frequency-domain harmonic model for compact fluorescent lamps. IEEE Transactions on Power Delivery. 2010; **25**(2):1182-1189

[8] Constantinescu F, Gheorghe AG, Marin ME, Taus O. Harmonic balance analysis of home appliances power networks. In: 2017 14th International Conference on Engineering of Modern Electric Systems (EMES). Oradea, Romania: IEEE; 1-2 June 2017. pp. 256-260

[9] Gheorghe AG, Constantinescu F, Marin ME, Ştefănescu V, Vătăşelu G. Frequency domain models for nonlinear home appliance devices. In: 2018 International Symposium on Fundamentals of Electrical Engineering (ISFEE). Bucharest, Romania: IEEE; 1-2 November 2018. pp. 1-4

[10] Constantinescu F, Gheorghe AG, Marin ME, Vătăşelu G, Ştefănescu V, Bodescu D. New models for frequency domain simulation of home appliances networks. In: 11-th International Symposium on Advanced Topics in Electrical Engineering. Bucharest, Romania: IEEE; 28-30 March 2019. pp. 1-5

[11] Technical University "Gheorghe Asachi", Iaşi, Romania, Faculty of Electrical Engineering and Power Engineering, Power Electronics Laboratory (in Romanian). Available from: http://www.euedia.tuiasi.ro/lab_ep/ep_files

[12] Gheorghe AG, Marin ME, Dragusin MD. New frequency domain models for multiple port devices with variable firing angle. In: 2021 25th International Conference on Circuits, Systems, Communications and Computers (CSCC), Crete Island, Greece. Vol. 2021. IEEE; 2021. pp. 51-55

[13] Constantinescu F, Rață M, Enache FR, Vătăşelu G, Ştefănescu V, Milici D, et al. Frequency domain models

for nonlinear loads with firing angle control devices part I – Measurements. In: 2019 15th International Conference on Engineering of Modern Electric Systems (ICEMES). Oradea, Romania: IEEE; 13-14 June 2019. pp. 241-246

[14] Enache A, Petrescu T, Enache FR, Popescu FG. Spectrum computation for periodic analog signals using genetic algorithms. In: ICATE 2014 - International Conference on Applied and Theoretical Electricity. Craiova, Romania: IEEE; 23-25 October 2014. pp. 1-6

[15] Enache FR, Constantinescu F, Rață M, Vătășelu G, Ștefănescu V, Milici D, et al. Frequency domain models for nonlinear loads with firing angle control devices, part II – Modeling. In: 6-th International Symposium on Electrical and Electronics Engineering. Galati, Romania: IEEE; 18-20 October 2019. pp. 1-6

[16] Gheorghe AG, Constantinescu F, Marin ME. Time domain and frequency domain models for analysis of nonlinear power networks of home appliances. In: International Symposium on Signals, Circuits and Systems 2021. Iasi, Romania: IEEE; 15-16 July 2021. pp. 1-4

[17] Lin S, Kuh ES, Marek-Sadowska M. Stepwise equivalent conductance circuit simulation technique. IEEE Transactions on Computer-Aided Design of Integrated Circuits and Systems. 1993; **12**(5):672-683

Chapter 4

Perspective Chapter: Mitigation of Power System Harmonics with the Incorporation of Active Filter for a Radial Distribution System

Nagaraj Ramrao, Busireddy Hemanth Kumar, Ponguleti Sandhya, Rangu Seshu Kumar and Arvind Singh

Abstract

Harmonics in power systems can cause various problems, including equipment damage, power quality degradation, and increased losses. Active filters have been proven to be an effective solution for mitigating harmonics in power systems. In this research work, the effectiveness of active filters in reducing harmonics is evaluated in MATLAB environment by implementing sparrow search optimization technique. To carry out the simulation results a standard IEEE 13 bus test system and unbalanced power system is considered to meet the IEEE 519 standards. The obtained simulation results demonstrate the significant reduction of harmonics with incorporation of active filters. The obtained simulation results show that hybrid active filter provides the best harmonic mitigation performance. The analysis shows that the use of active filters is economically feasible for reducing harmonics in power systems. Finally, this book chapter provides valuable insights into the application of active filters for power system harmonics mitigation and can help power system engineers and operators to improve the quality and reliability of their systems by implementing Sparrow search optimization technique.

Keywords: harmonics, sparrow search algorithm, active filter, IEEE 13 bus, power quality, distributed generation

1. Introduction

Electricity is provided via power systems, which are important infrastructures for a variety of industrial, commercial, and residential purposes. Power system harmonics, a major problem brought on by the growing usage of non-linear loads and power electronic devices, poses a serious obstacle. Harmonics are unwanted distortions in the voltage and current waveforms that can cause several problems, including higher losses, poorer power quality, and malfunctions in delicate equipment. As a result, power system engineers and researchers are now extremely

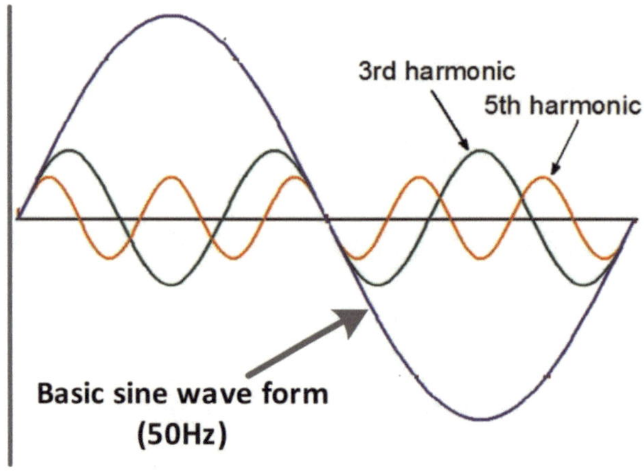

Figure 1.
Harmonic distortion waveforms of power system.

concerned with the reduction of power system harmonics [1]. Newfound options for addressing harmonics-related issues have emerged with the emergence of sophisticated power electronic devices and control techniques. To reduce harmonics and boost the performance of the entire power system, researchers have been concentrating on creating creative solutions that integrate hardware and software techniques. In terms of harmonics mitigation, the addition of renewable energy sources like solar and wind to the power grid has created new difficulties. Therefore, there is an urgent need for current research and useful techniques to handle these harmonics-related problems and guarantee the dependable and effective operation of power systems. The power grid is subject to various disturbances, including power system harmonics. Harmonics are voltage or current waveforms with frequencies that are integer multiples of the fundamental frequency, typically 50 Hz or 60 Hz. The general schematic view of harmonics waveform with respect to reference waveform is shown in **Figure 1** respectively.

These harmonics arise from non-linear loads, such as power electronic devices, and can cause a range of problems, including increased losses, reduced power factor, equipment overheating, and interference with communication systems [2]. In this book chapter the mitigation of power system harmonics for the IEEE 13 bus radial distributed test system with the incorporation of shunt active filter. The shunt active filters play a vital role in mitigating harmonic distortion, improving power quality, ensuring compliance with standards, compensating reactive power, and enhancing the performance and reliability of power systems. They are an essential tool in modern electrical networks where non-linear loads are prevalent.

2. Significance of power quality in distributed power system

Poor power quality can lead to a range of issues, particularly in microelectronics environments. In the past, electrical problems on mechanical equipment may have gone unnoticed, but they can significantly impact the operations of high-tech equipment. Understanding and preventing power-related issues is crucial for equipment owners, managers, designers, and other meter users, as a substantial portion of power quality problems arise from the consumer's end. Power quality pertains to the physical properties of the supplied power during regular conditions, ensuring that it does not cause any disruptions or problems in the customer's processes [3]. Voltage, current, or frequency deviations resulting from power quality problems can cause equipment failures or malfunctions for customers [4]. The quality of power supply focuses primarily on voltage profile improvement and the reliability of electrical supply. Voltage disturbances occur when the phase voltage deviates from its normal characteristics, potentially leading to meter malfunctions. A reliable electrical supply, on the other hand, is adequate (in that it can meet the demand), secure (in that it can survive unexpected problems like system malfunctions), and available (in that it can prevent long-term outages) [5, 6]. Disruptions in power quality are widespread in commercial, industrial, and utility networks. Lightning events often bring on these disruptions. Disruptions in the electrical grid can also be caused by switching events and oscillatory transients. Current and voltage harmonic components are generated when non-linear loads are applied. Harmonics are voltage and current sinusoidal waves whose frequency is an integer multiple of the fundamental frequency. These periodic voltages are added on top of the system's sinusoidal voltage. As a result, other devices plugged into the electrical system are subjected to increased strain due to these currents and voltages [7]. To overcome those concerns optimal placement of shunt active filters integrated with the radial distribution system is one of the key solutions. Shunt active filters play a significant role in reducing the total harmonic distortion (THD) in power systems. Shunt active filters are purpose-built to minimize the presence of harmonics within the power system. They achieve this by actively injecting harmonic currents that possess equal magnitudes but opposite phases to the existing harmonics. As a result, shunt active filters effectively cancel out these harmonics, leading to a reduction in total harmonic distortion (THD). This reduction in THD facilitates the generation of cleaner and more sinusoidal voltage and current waveforms. To maintain adherence to power quality standards and regulations like IEEE 519 and IEC 61000-3-4, power systems can employ shunt active filters to mitigate THD effectively.

3. Assessment of power quality indices under the harmonics

Power quality indices offer a quantitative assessment of the electrical power quality within a system. Harmonics, which refer to undesirable voltage or current distortions occurring at frequencies that are integer multiples of the fundamental frequency, can have a notable impact on power quality. The following are several commonly employed power quality indices utilized to evaluate the influence of harmonics [8].

Total Harmonic Distortion: The fundamental metric for measuring the degree to which distorted waveforms deviate from a pure sine wave is total harmonic distortion (THD). It provides a numerical representation of the power grid's current and voltage waveform distortion. The total harmonic distortion (THD) is calculated by dividing the sum of all harmonic components by the power of the fundamental frequency. Eq. (1) gives a the generic mathematical formulation of the THD.

$$THD = \sqrt{ID_1^2 + ID_2^2 + ID_3^2 \ldots \ldots \ldots \ldots \ldots ID_n^2} \tag{1}$$

Eq. (2) illustrates the mathematical representation of Total Harmonic Current (THC), which is caused by the summation of current orders from 2 to 40. The value of THC serves as the foundation for the installation of active filters. Mathematical representation of THC can be expressed as follows:

$$THC = \sqrt{\sum_{n=2}^{n=40} Ih^2} \tag{2}$$

Eq. (3) represents the mathematical expression of Total Harmonic Distortion Current (THD_i), which quantifies the level of distortion in a waveform. THD_i is derived by dividing the THC by the fundamental current. This equation illustrates the relationship between THD_i, THC, and the fundamental current.

$$THD_i = \sqrt{\sum_{n=2}^{n=40} \frac{Ih^2}{I_1}} = \frac{THC}{I_1} \tag{3}$$

I_1 represents fundamental current component and I_h represents harmonic current of the nth order. Eq. (4) represents the mathematical expression of Total Harmonic Distortion voltage (THD_v) respectively, which quantifies the level of distortion in a waveform. THD_V is derived by dividing the THC by the fundamental voltage. This equation illustrates the relationship between THD_v, THC, and the fundamental voltage.

$$THD_v = \sqrt{\sum_{n=2}^{n=40} \frac{{V_n}^2}{V_1}} = \frac{THC}{V_1} \tag{4}$$

Shunt active filters serve as electronic devices employed in power systems to mitigate harmonic distortion and enhance power quality. Their specific purpose is to compensate for reactive power and harmonic currents within the system, resulting in the reduction of voltage distortion and improvement of the overall power factor. $\boldsymbol{V_1}$ indicates fundamental voltage component and $\boldsymbol{V_n}$ represents voltage of nth harmonic order.

3.1 Integration of active filters in distribution systems

The integration of hybrid active filters into distributed systems involves incorporating these filters into power distribution networks. The primary objective is to address power quality concerns arising from harmonics, voltage fluctuations, and other disturbances. Hybrid active filters leverage the benefits of both active and passive filters to deliver efficient and reliable compensation for power quality issues.. The primary aim of active filtering is to address these issues dynamically, instead of relying on predetermined components with high ratings, which are typically bulky passive components [9]. This approach allows for a significant reduction in rating requirements. Based on the specific nature of the problem, active filters can be implemented in three primary topologies: shunt type, series type, or a combination of both known as shunt-series type active filters [10]. The schematic representation of

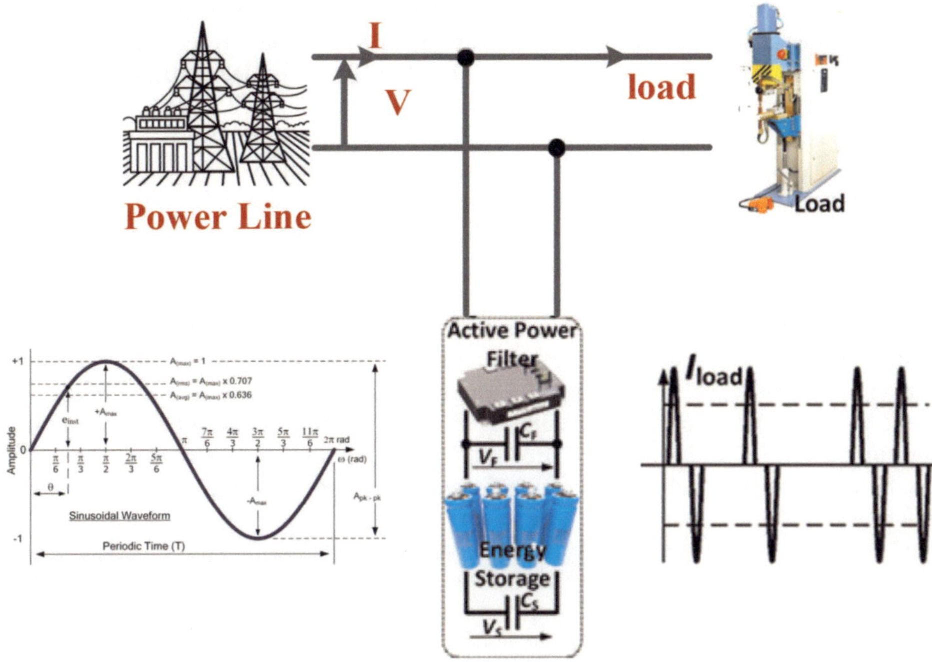

Figure 2.
Integration of active filter with distributed test system.

active filter is integrated with distributed system is shown in **Figure 2** respectively. By integrating hybrid active filters into distributed systems, several benefits can be achieved. Hybrid active filters exhibit effective harmonic suppression, reducing distortion levels and ensuring power supply quality remains within acceptable limits. They provide voltage regulation capabilities to compensate for sags, swells, and fluctuations, thereby ensuring a stable and reliable power supply. With fast response times and adaptability to changing system conditions, hybrid active filters are well-suited for distributed systems that experience diverse loads and disturbances. Moreover, the active filters integrated into the hybrid configuration enhance power factor, compensating for reactive power, minimizing energy losses, and improving overall system efficiency.

3.2 Active filter design

The Active Harmonic Filter, utilizing IGBT semiconductors and multiple control loops, injects a dynamic cancelation signal into the power line, effectively reducing harmonics and improving Power Factor. This advanced technology from Power Correction Systems enhances AC Motor Systems and AC Variable Frequency Drive (VFD) Systems' performance and functionality, while seamlessly integrating with a wide range of electrical and electronic devices in their ability. Active filters provide harmonic compensation without concerns about reactive power at fundamental frequencies. Consequently, the rated power of an active filter can be lower compared to an equivalent passive filter serving the same non-linear load. Moreover, active filters avoid causing system resonances that might otherwise shift a harmonic problem from one frequency to another. Power electronics play a crucial role in the active filter concept by generating harmonic current components that counteract those from non-linear loads. The active filter achieves this by utilizing power electronic switching to nullify the harmonic currents produced by the non-linear load. The foundation of the active filter architecture under study in this lecture is a pulse-width modulated (PWM) voltage source inverter, which connects to the system through a system interface filter, as depicted in **Figure 3** respectively. In this setup, the filter and the load being corrected are connected in parallel, commonly referred to as an active parallel or shunt filter.

The SAPF structure comprises two main components: power circuits, consisting of power semiconductor switches, capacitors, inductors, and possibly power diodes in certain SAPF topologies, and the control system, designed to regulate the switching function of the switches. Using **Figure 3** and the PCC, one can apply Kirchhoff's current law (KCL) to calculate the current flow in a harmonic-polluted power system before integrating a Active Power Filter (APF). This makes understanding the fundamental concept of APF relatively straightforward through observation of current flow in the power system.

$$i_s = i_L = i_{1L} + i_H \tag{5}$$

In the context of i_S representing the source current and i_L representing the load current, the latter may consist of i_{1L} (fundamental current) and i_H (harmonic current caused by the presence of harmonic-producing loads). It is crucial to recognize that i_S is currently distorted and not in phase with the source voltage V_s, primarily due to the influence of i_H. Therefore, the main objective of implementing a shunt active power filter is to eliminate i_H.

Figure 3 illustrates the addition of two more current flows to the power system following the installation of SAPF at PCC. Firstly, the SAPF injects a mitigation current into the power system through PCC to cancel out i_H (referred to as the

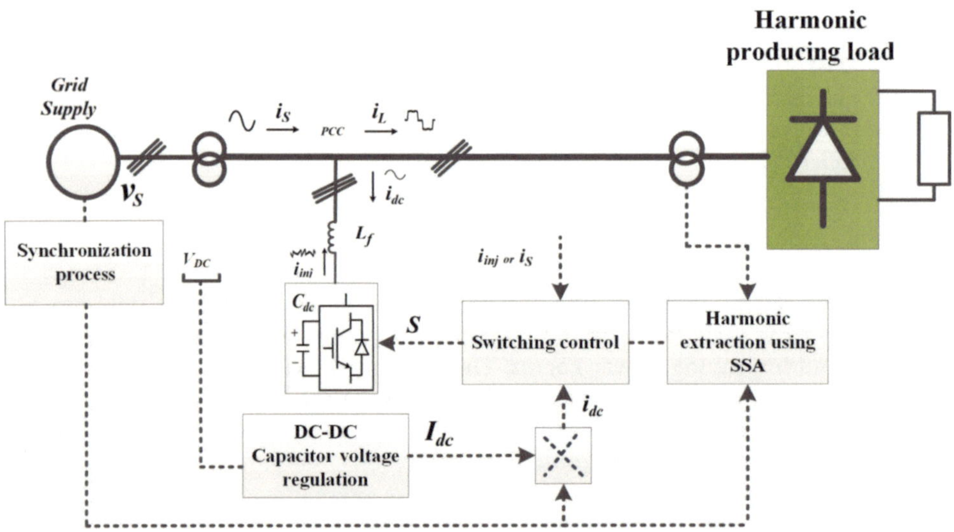

Figure 3.
Active filter design and methodology.

injection current in this study). The injected mitigation current matches the size and phase of the i_H current. Secondly, the SAPF utilizes a small amount of current (known as the dc-link charging current i_{dc}) to maintain a constant voltage V_{dc} across its dc-link capacitor C_{dc}. This ensures effective control of switching losses and ensures the SAPF operates consistently and reliably. Eq. (5) can be expressed mathematically using KCL as follows;

$$i_s = [i_L + i_H] - i_{inj} + i_{dc} \tag{6}$$

The voltage level across the dc-link capacitor has a direct impact on the size of the produced i_{inj}. The produced i_{inj} will perfectly match thei$_H$ current, causing their total cancelation, after the voltage across the dc-link capacitor has reached the correct level and is constantly maintained. As a result, Eq. (6) can be further simplified and represented in Eq. (7) respectively.

$$i_s = i_{1L} + i_{dc} \tag{7}$$

With this regard the harmonic distortion of the implemented test system has mitigated effectively by incorporating shunt active power filters and improves the system stability.

4. Optimization techniques taxonomy in reduction of THD

Optimization is a fundamental mathematical discipline extensively utilized across various engineering fields. It provides essential tools in the pursuit of increasingly optimal solutions. Typically, an optimization problem encompasses a defined objective function and constraints. In many scenarios, multiple objectives exist simultaneously, necessitating the use of several goal functions. To attain the best possible outcome, a multicriteria analysis becomes imperative. Multi-objective optimization is notably complex since objectives often conflict with each other, requiring a trade-off to be established. As a result, solutions obtained through multi-objective optimization represent Evolution-based algorithms, inspired by natural evolution, and are utilized to generate populations for algorithmic solutions [11, 12]. These algorithms involve creating individuals through processes such as mutation, crossover, or selection of the best solutions from a mathematical model [13]. The Genetic Algorithm (GA) is a well-known example of this type of algorithm because it is inspired by Darwin's theories of evolution. Differential Evolution (DE), Backtracking Search Algorithm (BSA), and the Evolution Strategy are only some of the numerous methods that have been created.

Algorithms based on Swarm Intelligence mimic the cooperative efforts of insects, fish, and birds as they forage for food or pursue prey. These collective actions serve as the basis for mathematical models [14, 15]. Particle swarm optimization (PSO), created by Kennedy and Eberhart, is a well-known example of such an algorithm. Cat swarm optimization (CSO) is another computational paradigm like those used by ants and honeybees. Algorithmic techniques like Simulated Annealing (SA) and the Gravitational Search Algorithm (GSA) are based on the principles of physics that govern the cosmos [16]. Modeling human behavior mathematically allows for the development of relational algorithms. The success of these models has a one-to-one correlation with how people act [17, 18]. These trade-off solutions, which enhance one

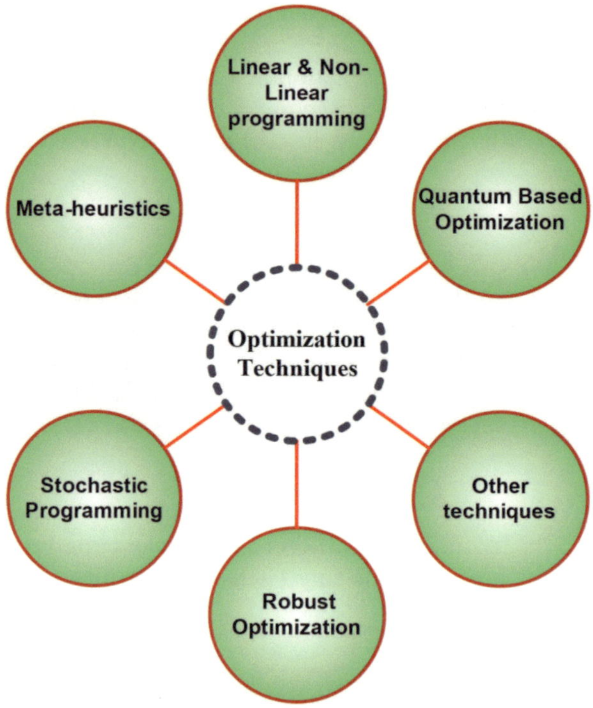

Figure 4.
Optimization taxonomy.

criterion while sacrificing another, constitute a Pareto set. The selection of a solution from the Pareto set ultimately depends on individual preferences. So far many of the research problems have been solved by implementing PSO, GA, and other mathematical algorithms. The possibility of implementing Sparrow search algorithm in harmonic mitigation of IEEE- 13 bus test system is identified in this book chapter. One of the applications of sparrow search algorithm is implemented to solve optimal energy management of grid-connected microgrid problem in [19]. The optimization taxonomy which includes heuristic, meta-heuristic and other mathematical optimization approaches is represented in **Figure 4** respectively.

4.1 Methodology for minimizing THD of IEEE: 13 bus test system

The use of optimization algorithms influenced by nature has been increasingly common in recent decades [20] due to the tendency of deterministic algorithms to get stuck in local optima. Swarm intelligence-based optimization algorithms are at the forefront of the field because of their ability to efficiently solve global optimization problems. The Sparrow Search Algorithm, developed by Jiankai Xue and Bo Shen [21] is a notable addition to this class. The biological traits of sparrows inspired this program's robustness and stability. The system took cues from the birds' foraging behavior, collective knowledge, and anti-predator strategies. Sparrows, intelligent omnivores that primarily feed on grains and weeds, are used as a basis for the mathematical modeling of SSA. Both producers and scavengers, these sparrows use a wide variety of foraging strategies to ensure the survival of their flocks. Scroungers rely on producers to get their food, while producers actively search for it. Seventy-five percent of the sparrow hosts act as producers, while the other 25% act as scavengers. Foraging refers to the practice of gathering food in a social group. When one or a few sparrows sneak away with food while evading predators, they sound an alarm chirp to warn their fellow birds. **Figure 5** depicts how sparrows forage for food. The mathematical formulation of the sparrow optimization approach and its applications in solving complex engineering problems is briefly discussed in [22] respectively. In [23] optimal placement of DG unit problem is solved by implementing intelligent optimization approach. The harmonic mitigation problem is tackled in [24] by implementing ETAP software. So far discussed in the existing literature review the optimization approaches are implemented for different research problems like energy management, optimal scheduling, overall reduction of operating costs respectively. But in case of mitigation of harmonic distortion of the IEEE-13 bus test system by using sparrow search algorithm is not implemented. The application of sparrow search optimization approaches has been discussed in [25] with real time engineering problems. The proposed algorithm is to evaluate the performance of IEEE-13 bus radial distributed test system followed by minimizing harmonic distortion content. The decision variables are voltage and current harmonics followed by Eq. [3] and Eq. [4] respectively. The proposed problem is deal with the harmonic mitigation of the IEEE-13 bus standard test system. The proposed SSA algorithm is implemented to find the distribution network load flow parameters such as voltage profile at each node and other network parameters subjected to load flow analysis with and without incorporation of active filters. Further the incorporation of shunt active filter in the test

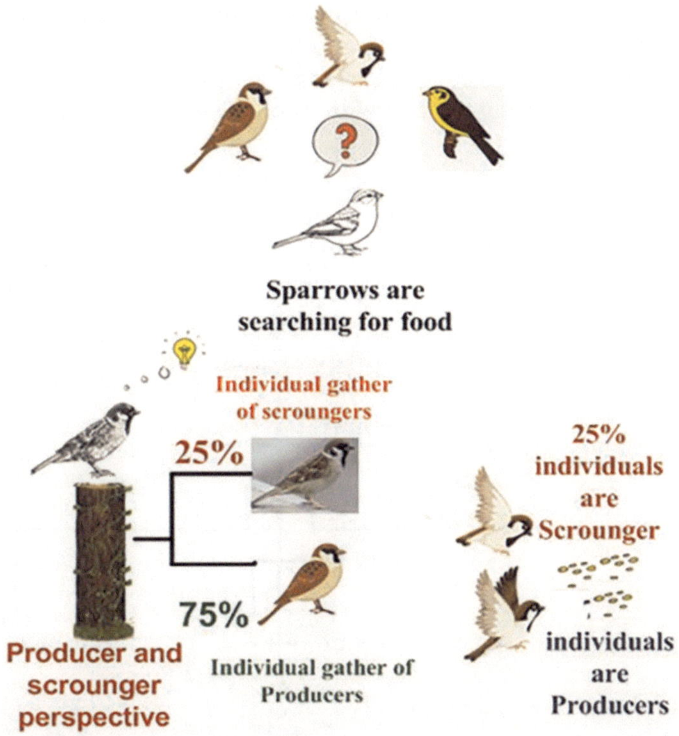

Figure 5.
Conceptual illustration of foraging behavior of sparrows [22].

system is evaluated all the network flow parameters by using SSA algorithm. The obtained simulation results are represented by using FFT window in MATLAB to analyze the harmonic distortion for the IEEE 13 bus test system respectively.

5. Numerical evaluation and discussion

In this book chapter mitigation of power system harmonics with the integration of active filters by implementing sparrow search optimization technique is evaluated. The considered standard radial distributed test system IEEE 13 bus with the integration of active filter is optimally placed to enhance the voltage profile indexes and mitigate the harmonic distortion content in the considered test system. The single line diagram of IEEE 13 bus radial distributed test system is shown in **Figure 6** respectively. The proposed optimizer sparrow search effectively mitigates the harmonic content distortion of voltage and current of the considered test system. The incorporation of active filters with the unbalanced test system mitigates the harmonic content distortion and enhances the voltage profile indices effectively are represented in **Figures 7** and **8** respectively.

The primary objective of this book chapter is implementing sparrow search algorithm on IEEE 13-bus radial distributed unbalanced test system equipped with active filters is evaluated within the MATLAB computing environment. The implementation was conducted on an HP Laptop equipped with an Intel Core i7 processor running at 2.4 GHz and with a RAM capacity of 12 GB. The effectiveness of the developed algorithm was evaluated using an unbalanced-13-bus radial distribution system (unbalanced-13-bus-RDS).A harmonics analysis is performed on an IEEE 13-bus distribution system that supplies various types of industrial and commercial loads, as shown in **Figure 5**. The system comprises a main supply at 69 kV connected to bus 4 and a local generator operating at bus 1 with a voltage of 13.8 kV. A 6000 kVAr

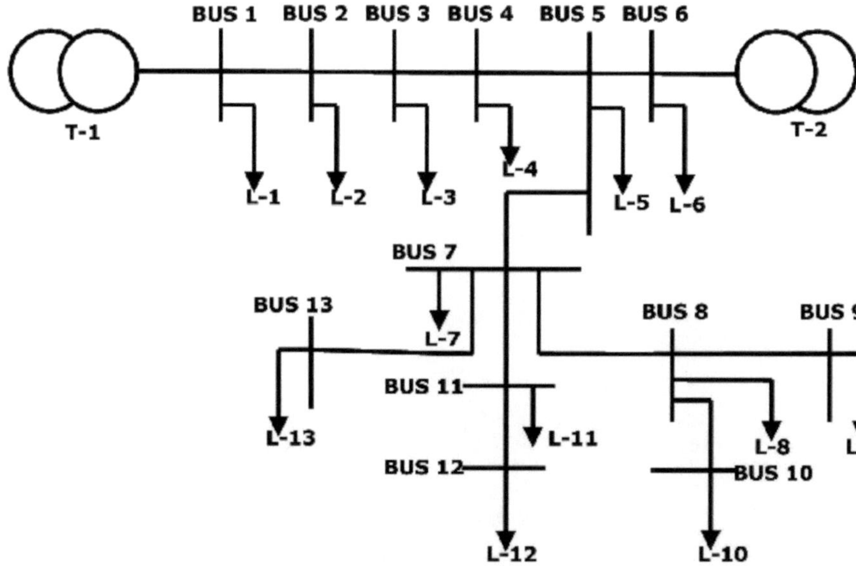

Figure 6.
Single line diagram of IEEE-13bus radial distributed test system [23].

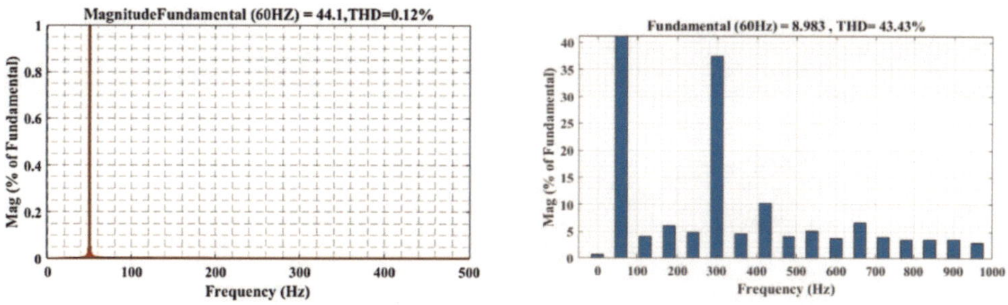

(a) (b)

Figure 7.
Voltage THD of bus number 10 (a) with active filter (b) without filter.

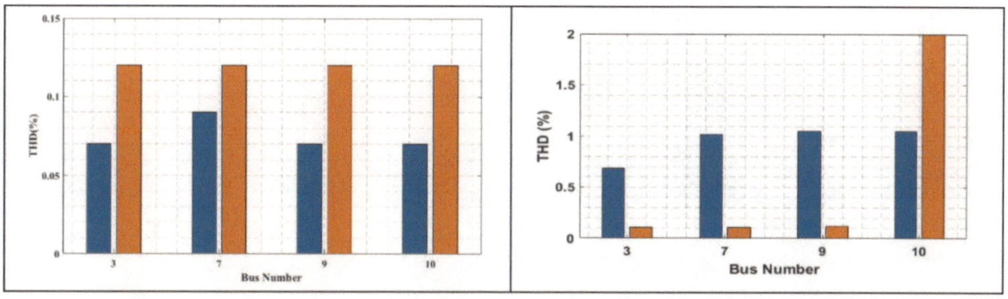

(a) (b)

Figure 8.
Total harmonic distortion (%) of the corresponding bus number (a) with active filter (b) without filter.

capacitor, rated for improving power factor, is connected at bus 3. The customers on bus 7 and bus 10 are served by non-linear loads, which are known to produce harmonics. The integration of active filters to the IEEE-13bus distributed test system with and without presence the voltage profiles and total harmonic distortion is shown in **Figures 7** and **8** respectively.

To conduct further testing on the proposed sparrow search algorithm, the active filter incorporation in the power system for reducing total harmonic distortion content which includes voltage and current harmonics. Instead of placement of DG units or harmonic filters the shunt active hybrid filters play a significant role for reducing power system harmonics. To determine the optimal settings of the proposed algorithm parameters, 20 independent runs were performed, with a maximum of 100 iterations allocated for adjusting each parameter. A swarm population size of 50 particles, which includes number of Producers are 0.8, and number of Scroungers are 0.2 respectively. By utilizing the developed sparrow search algorithm, the optimal location of active filters in the considered test system were successfully identified, resulting in minimized overall total harmonic distortion of the IEEE-13 bus radial distributed unbalanced test system. In addition to the evaluation of total harmonic distortion of IEEE-13 bus distributed test system with the proposed algorithm the voltage profiles of the corresponding busses are evaluated and represented in **Figure 9** respectively. The voltage profile at bus 10 has been enhanced by 7% after the incorporation of active

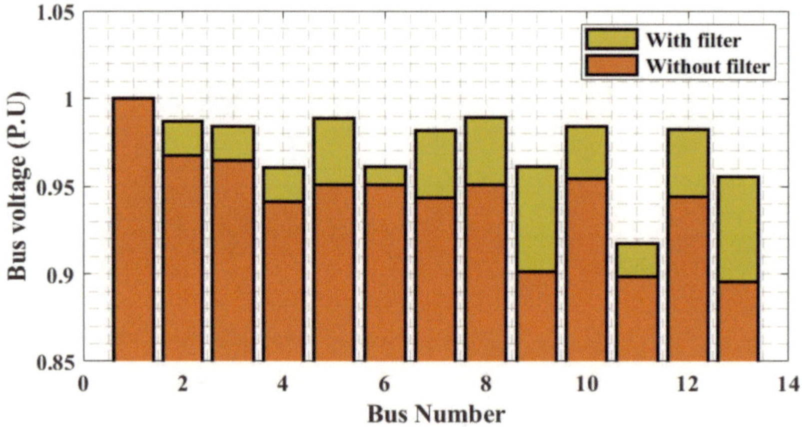

Figure 9.
Voltage profile with and without incorporation of active filters.

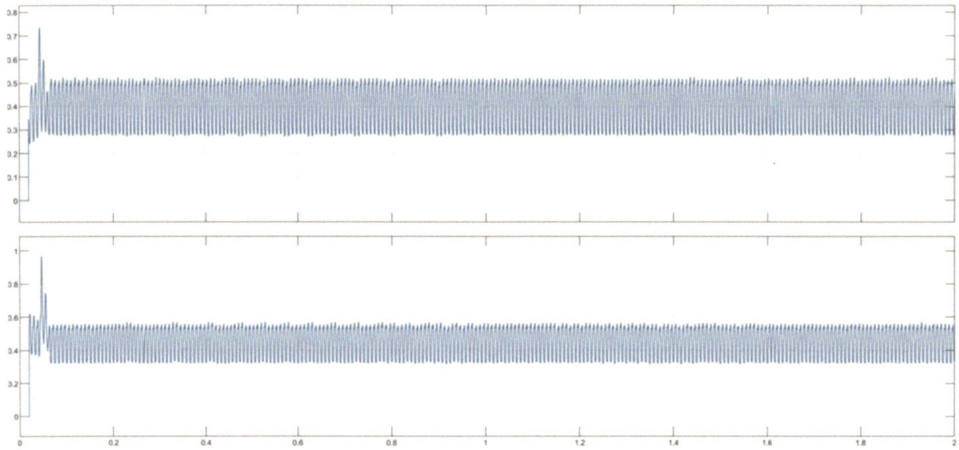

Figure 10.
Current harmonics with and without incorporation of filter.

power filter. The incorporation of active power filters in the considered test system enhances the voltage profiles of the corresponding busses and improves the system stability. **Figures** 7 and **8** illustrate that the incorporation of active filters in the 13-Bus-RDS results in an improvement in the voltage profile. The current harmonics of the considered unbalanced test system with and without incorporation of active filter is shown in **Figure 10** respectively. Conversely, by considering harmonics, the harmonic distortion levels at load busses are kept within permissible limits according to the IEEE 519 standards respectively.

6. Conclusion

In this book chapter mitigation of total harmonic distortion of IEEE- 13 bus radial distributed standard test system with the incorporation of active filters by implementing sparrow search optimization technique respectively. The main objective of this book chapter is to mitigate the harmonic distortion content of the considered test system associated with non-linear loads. The voltage profiles have been enhanced by incorporating active filters that brings the system into a stabilizer manner. The voltage profiles at the corresponding busses are enhanced effectively by incorporating active filters. The findings of the study show that ignoring the existence of harmonics and incorporating hybrid active filters in the system can result in unwanted levels of harmonic distortion, resulting in greater damage to electrical equipment for both the electric utility and consumers.

Author details

Nagaraj Ramrao[1*], Busireddy Hemanth Kumar[1*], Ponguleti Sandhya[2], Rangu Seshu Kumar[3] and Arvind Singh[4]

1 School of Engineering, Mohan Babu University, Tirupati, India

2 Department of EEE, The Oxford College of Engineering, Bengaluru, India

3 KPR Institute of Engineering and Technology, Coimbatore, India

4 University of Pretoria, South Africa

*Address all correspondence to: nagaraj.ramrao@gmail.com and hemub09@gmail.com

References

[1] Venkatesh C, Kumar D, Siva Sarma D, Sydulu M. Estimation and mitigation of voltage and current harmonics in distribution system. In: TENCON 2008 – 2008 IEEE Region 10 Conferences. Hyderabad, India; 2008. pp. 1-6

[2] RaviKumar G, Lokya M, Vijay Muni T. Estimation and minimization of harmonics in IEEE 13 bus distribution system. International Journal of Engineering Research & Technology (IJERT). 2012;**1**(7):1-6

[3] Khan H, Choudhry MA, Mahmood T, Hanif A. Investigating the electric power distribution system (EPDS) Bus Voltage in the Presence of Distributed Generation (DG). In: Proceedings of the 5th WSEAS International Conferences on Instrumentation, Measurement, Circuits and Systems, April 16–18, 2006. Hangzhou, China; 2006. pp. 207-212

[4] Moreno-Munoz A et al. Improvement of power quality using distributed generation. ELSEVIER Journal on Electrical Power and Energy Systems. 2010;**32**:1069-1076

[5] Salam MA. —Power quality disturbances in a test distribution system: An overview. Journal of Applied Sciences Research. 2013;**9**(1): 560-566

[6] Enslin JHR, Heskes PJM. — Harmonics interaction between a large number of distributed power inverters and the distribution network. IEEE Transaction Power Electronics. 2004; **19**(6):1586-1593

[7] Carrasco M, Franquelo LG, Bialasiewicz JT, Galvan E, Guisado RCP, Prats MAM, et al. Power electronic systems for the grid integration of renewable energy sources: A survey. IEEE Transaction Industrial Electronics. 2006;**53**(5):1398-1409

[8] Okojie DE, Omoregbee HO, Richards CG, et al. Harmonic generation of variable speed drive under complex-voltage unbalance conditions. Journal of Engineering and Applied Science. 2022; **69**:96, 1-16. DOI: 10.1186/s44147-022-00146-9

[9] Akagi H. Modern active filters and traditional passive filters. Bulletin of the Polish Academy of Sciences, Technical Sciences. 2006;**54**(3):255-269

[10] Salam Z, Cheng TP, Jusoh A. Harmonics mitigation using active power filter: A technological review. In: ELEKTRIKA. Vol. 8, No. 2. Malaysia. pp. 17-26

[11] Baros J, Sotola V, Bilik P, Martinek R, Jaros R, Danys L, et al. Review of fundamental active current extraction techniques for SAPF. Sensors. 2022;**22**:7985, 1-41. DOI: 10.3390/s22207985

[12] Martinek R, Bilik P, Baros J, Brablik J, Kahankova R, Jaros R, et al. Design of a measuring system for electricity quality monitoring within the SMART street lighting test polygon: Pilot study on adaptive current control strategy for three-phase shunt active power filters. Sensors. 2020;**20**:1718, 1-31. DOI: 10.3390/s20061718

[13] Bilal PM, Zaheer H, Garcia-Hernandez L, Abraham A. Differential evolution: A review of more than two decades of research. Engineering Applications of Artificial Intelligence. 2020;**90**:103479. DOI: 10.1016/j.engappai.2020.103479

[14] Jesús GF, Carlos A. Indicator-based multi-objective evolutionary algorithms: A comprehensive survey. ACM Computing Surveys. 2020;**53**:1-35. DOI: 10.1145/3376916

[15] Rostami M, Berahmand K, Nasiri E, Forouzandeh S. Review of swarm intelligence-based feature selection methods. Engineering Applications of Artificial Intelligence. 2021;**100**: 104210. DOI: 10.1016/j.engappai. 2021.104210

[16] Tang J, Liu G, Pan Q. A review on representative swarm intelligence algorithms for solving optimization problems: Applications and trends. IEEE/CAA Journal of Autonomous. 2021; **8**:1627-1643. DOI: 10.1109/jas.2021. 1004129

[17] Shashank RV, Sai NB, John CM, Elizabeth M. A review of physics-based machine learning in civil engineering. Results Engineering. 2022;**13**:100316, 1-12. DOI: 10.1016/j.rineng.2021.100316

[18] Burton JW, Stein M, Jensen TB. A systematic review of algorithm aversion in augmented decision making. Journal of Behavioral Decision Making. 2019;**33**: 220-239. DOI: 10.1002/bdm.2155

[19] Raghav LP, Rangu SK, Dhenuvakonda KR, Singh AR. Optimal energy management of microgrids-integrated nonconvex distributed generating units with load dynamics. International Journal of Energy Research. 2021;**45**(13):18919-18934. DOI: 10.1002/er.6995

[20] Lolla PR, Rangu SK, Dhenuvakonda KR, Singh AR. A comprehensive review of soft computing algorithms for optimal generation scheduling. International Journal of Energy Research. 2020;**2020**:1-20. DOI: 10.1002/er.5759

[21] Xue J, Shen B. A novel swarm intelligence optimization approach: Sparrow search algorithm. Systems Science & Control Engineering. 2020; **8**(1):22-34. DOI: 10.1080/ 21642583.2019.1708830

[22] Singh AR, Ding L, Raju DK, Raghav LP, Kumar RS. A swarm intelligence approach for energy management of grid-connected microgrids with flexible load demand response. International Journal of Energy Research. 2022;**46**(4):4301-4319. DOI: 10.1002/er.7427

[23] Aref A, Davoudi M, Razavi F, Davoodi M. Optimal DG placement in distribution networks using intelligent systems. Energy and Power Engineering. 2012;**4**:92-98. DOI: 10.4236/epe.2012. 42013

[24] Mahiwal LG, Jamnani JG. Analysis and mitigation of harmonics for Standard IEEE 13 bus test system using ETAP. In: International Conference on Computing, Power, and Communication Technologies (GUCON). New Delhi, India; 2019. pp. 546-550

[25] Awadallah MA, Al-Betar MA, Doush IA, et al. Recent versions and applications of sparrow search algorithm. Achieves in Computational Methods Engineering. 2023;**30**: 2831-2858. DOI: 10.1007/s11831-023-09887-z

Chapter 5

Investigation on the Performance Efficiency of the Shunt Hybrid Active Power Filter

Chamberlin Stéphane Azebaze Mboving and Zbigniew Hanzelka

Abstract

The disturbances observed in the electrical system such as harmonics, asymmetry, flickers, voltage dips and swells, transients, are due to the type of connected devices. The nonlinear devices are the most common ones. Their connection to the electrical system without solution of mitigating their negative influence on supply network can be the cause of poor power quality. There are different types of solutions such as the passive harmonic filters (PHF), the active power filters (APF), the hybrid active power filters (HAPF). Each of those solutions presents some advantages and disadvantages. The chapter is focused on the HAPF, which is the combination of the PHF and shunt active power filter (SAPF) connected in series. Such topology is known in the literature, but still, there are certain problems that need more clarification through detailed studies. For instance, there are not enough recommendations on how to choose the PHF tuning frequency, when it is connected in series with SAPF. The investigations presented in this chapter are very detailed and based on a case study. The proposed control system algorithms of filters (SAPF and HAPF) are presented as well as formulated recommendations.

Keywords: passive, active and hybrid harmonic filter, harmonics, current unbalance, rate of current change, reactive power, switching ripples, control system

1. Introduction

In today societies, the production of nonlinear loads such as household appliances and industries electrical devices is in full grow. Their mass connection to the supply network (despite their compliance with standards) may cause a deterioration of the power quality.

The power quality refers mostly to the supply voltage quality (frequency, amplitude, waveform, unbalance, etc.) which should be in accordance with the recommendations set by the national or international standards (e.g., EN 50160 [1]).

If the supply voltage at the point of common coupling (PCC) presents poor quality (not complying standards), its improvement is therefore necessary. The poor power quality does not come from the energy producer (because the voltage at the terminal of power plants is almost without disturbances), but mainly from connected disturbing devices such as power electronic devices (e.g., diode and thyristor bridges), arc devices (e.g., arc furnaces, welding devices, discharge lamps), saturated magnetic cores devices (e.g., transformers, motors).

The electrical power is as a commodity and taking care of its quality is necessary. The power quality disturbances are numerous and varied (e.g., voltage drops and swells, flickers, harmonics, asymmetry) and their presence in the electrical system has consequences on the connected devices. For instance, the harmonics, if not mitigated can cause the increase of current RMS value; the overloading, overheating, and even damage of power system elements (e.g., transformers, generators, cables, electric motors, capacitors) and other connected devices (e.g., household appliances); the reduction of devices life span; the perturbation of the devices normal operation and power system operating costs increase; the inaccurate measurements of energy and power; decrease of power factor (PF), etc. [2–4].

To maintain the grid power quality in compliance with the standards, many solutions are proposed, including passive harmonic filters (PHF), active power filter (APF), hybrid power harmonic filter (HPHF), etc. [5]. Each of the solution presents advantages and disadvantages. The PHFs are applied in most cases for harmonics filtration and fundamental reactive power mitigation [6]. They present disadvantages such as: fixed reactive power and fixed tuning frequencies (e.g., designed for a defined load parameters), grid parameter dependency (e.g., grid impedance of the harmonic to be eliminated), resonance (series and parallel) problem, sensitive to the tolerance of its elements (e.g., reactor inductance), detuning phenomenon because of the aging, power losses in the case of filters with damping resistance (e.g., broad-band PHFs). Despite their drawbacks, the PHFs are still applied in practice and from the economical point of view, they are more preferred than the shunt active power filters (SAPFs) [7–10].

In comparison with the PHFs, the SAPFs are more efficient and their application is growing in low and medium voltage systems, particularly in the industries [11, 12]. The goal of their application can be the mitigation of disturbances such as current/voltage harmonics, asymmetry, and reactive power compensation (fundamental component). Their disadvantages are high cost [13], complex control system, difficulty for large-scale implementation [14, 15].

The hybrid filter topologies are diversified in the literature [16]. They result from the combination of the shunt or series APF with the parallel PHF or the shunt APF with the series APF. The main advantages of combining the active and passive filters together are as follows: (a) The power demand and performance cost of the active part are less than when it is operating alone and (b) the overcoming of the passive part disadvantages (e.g., resonance phenomena, grid impedance dependency) [17, 18].

The shunt hybrid active power filter (SHAPF) under study in the chapter is a structure known in the literature. It is composed of PHF connected in series with the SAPF (**Figure 1**) [19, 20]. The SAPF when operating alone is always connected to the full-supply phase-to-phase PCC voltage and need for its proper operation, a high rate of inverter DC voltage (e.g., 710 V and more [21, 22]). But connected in series with the PHF, the passive part allows the active part to work under a small rate of inverter voltage, therefore reducing its initial and operating cost [23–26]. The investigated SHAPF is for the grid current harmonics and fundamental harmonic reactive power compensation and its DC voltage is fixed at 150 V and can be less (e.g., 70 V (see [27])).

In the literature, there are not many clarifications on how to choose the tuning frequency of the PHF when it is connected in series with the inverter (SAPF). In Refs.

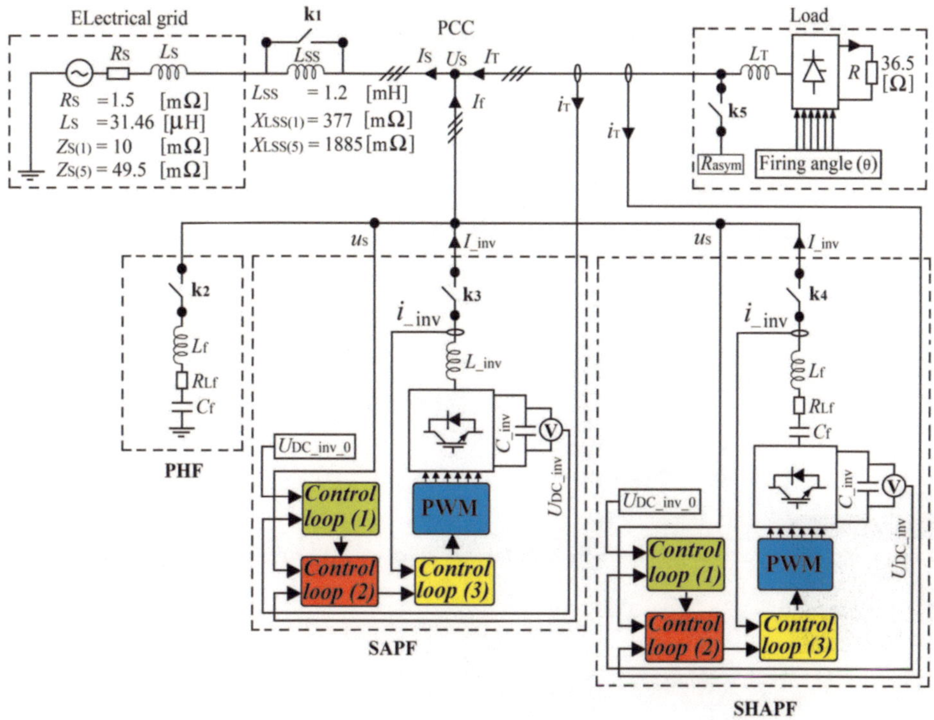

Figure 1.
Electrical system together with the investigated filters (k1-k5—Switches).

[28–30] for instance, it is recommended to tune it to the 7th harmonic frequency because it is then less bulk, low cost and presents lower impedance for the 11th and 13th harmonic frequencies than in the case when it is tuned to the 5th harmonic frequency. It is difficult to find papers presenting the investigations in which the SHAPF work efficiency is compared after tuning the passive part to different frequency. In this chapter, such investigations are proposed.

The chapter is organized in five parts: The first one presents the electrical system (grid and load) in which the analyses are performed; the second part presents the influence of the grid parameters on the PHF performance efficiency. In the third part, the SAPF with input reactor is investigated. In fourth one, the SHAPF with proposed control system is considered. The last part contains the conclusion. All the chapter investigations are performed in MATLAB/SIMULINK [31].

2. Electrical system description

The electrical system environment in which the filters are investigated is presented in **Figures 1** and **2**. The electrical grid equivalent parameters in **Figure 2** are taken from the laboratory setup.

In **Figure 3(a)** and **(b)**, it can be seen the PCC voltage waveforms together with the spectrum when the filters and load are not connected. The load is presented in **Figure 1**. It constitutes of six-pulse thyristor bridge with resistance (R) at DC side and input reactor (L_T). The phase-to-phase connected resistance R_{asym} (k5 closed) is used

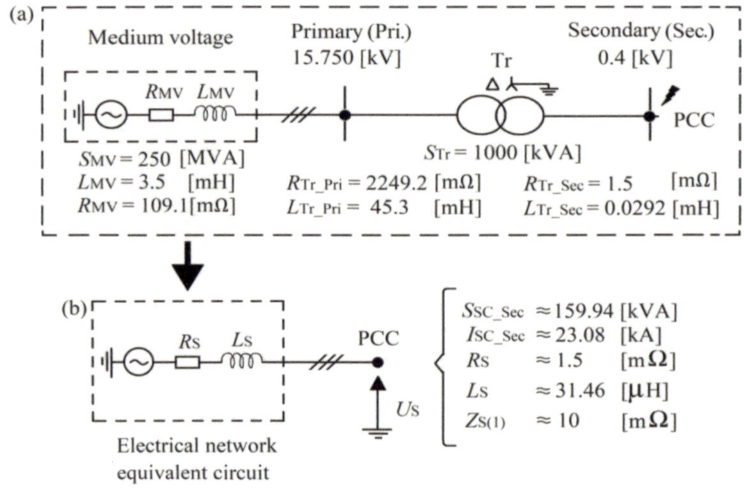

Figure 2.
(a) Parameters of the electrical grid taken from the laboratory set up, (b) electrical grid equivalent circuit.

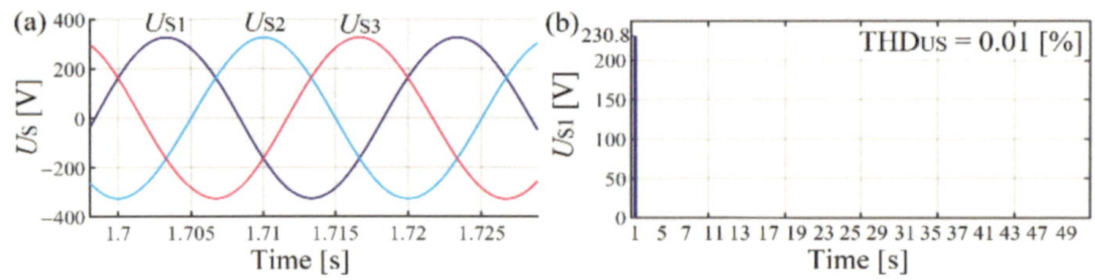

Figure 3.
(a) PCC voltage waveforms without any load or filter connected with (b) the spectrum (one-phase representation because of the symmetrical system).

to create load asymmetry in the electrical system in the investigation frame of the SAPF and SHAPF.

The additional line reactor L_{SS} (see **Figure 1**) connected between the grid and the PCC will be used in the investigations case of the PHF. Other parameters (e.g. L_T, filters) of the electrical system in **Figure 1** will be presented later.

3. Investigation on the grid dependency of the PHF work efficiency

This chapter investigates and compares two case studies: In the first one, the PHF is connected to the electrical system where the grid equivalent impedance for the harmonic to be eliminated ($Z_{S(5)}$) is smaller than the one of the filter ($Z_{f(5)}$). In the second one, the PHF is connected to the electrical system where the grid equivalent impedance for the harmonic to be eliminated ($Z_{S(5)}$) is increased by the additional line rector (L_{SS}) and is therefore higher than the one of the filter ($Z_{f(5)}$).

3.1 Simulation assumptions

The simulations are performed based on the assumptions that: the supply voltage before the PHF and the load connection does not contain harmonic (see **Figure 3**), the resistance of the additional line reactor (L_{SS}) as well as the one of the thyristor bridge input reactor (L_T) are neglected, the switches k3, k4, and k5 are opened and k2 is close (**Figure 1**). The switch k1 will be opened and closed depending on the case study. The electrical grid equivalent impedance for the 5th harmonic ($Z_{S(5)}$—**Figure 1**) is computed considering the electrical system parameters from the medium voltage side of the transformer (see **Figure 2(a)**). The 5th harmonic is the lowest characteristic harmonic generated by the load (after the fundamental) according to the formula $6k \pm 1$ wehre $k > 0$. Because of the symmetrical system, the results are presented only for one-phase.

3.2 Design of the PHF

The formula used to compute the PHF parameters is presented in **Figure 4(a)** and its impedance versus frequency characteristic is shown in **Figure 4(b)**. Because of the aging, the PHF is tuned to the frequency of 243.5 Hz ($n_{re} = 4.87$) which is a bit lower than the frequency of harmonic to be eliminated (5th). The computed parameters are shown in **Table 1**. Comparing the 5th harmonic equivalent impedance of the PHF ($Z_{f(5)} = 490$ [mΩ]) to the one of the grid ($Z_{S(5)} = 49.5$ [mΩ], when L_{SS} is not considered). It can be notices that (see **Table 1**) the one of the PHF is almost 10 times higher than the one of the grid, which allows to conclude that a big part of the 5th harmonic current coming from the load will not be filtered. That situation is contrary when the addition line reactor L_{SS} is considered ($Z_{S(5)} = 1934$ [mΩ] is almost 4 times higher than $Z_{f(5)}$).

For $R_{Lf} = R_{Cf} = 0$

$$\underline{Z}_f(j\omega_{(1)}) = j\frac{\omega_{(1)}^2 - \omega_{re}^2}{\omega_{re}^2\omega_{(1)}C_f} = j\frac{1 - n_{re}^2}{n_{re}^2\omega_{(1)}C_f}$$

$$C_f = \frac{n_{re}^2 - 1}{n_{re}^2}\frac{Q_f}{\omega_{(1)}U_f^2}$$

$$L_f = \frac{1}{n_{re}^2 - 1}\frac{U_f^2}{\omega_{(1)}Q_f}$$

$$q_{Lf} = \frac{X_{Lf}}{R_{Lf}}$$

$$n_{re}^2 > 1$$

(a)

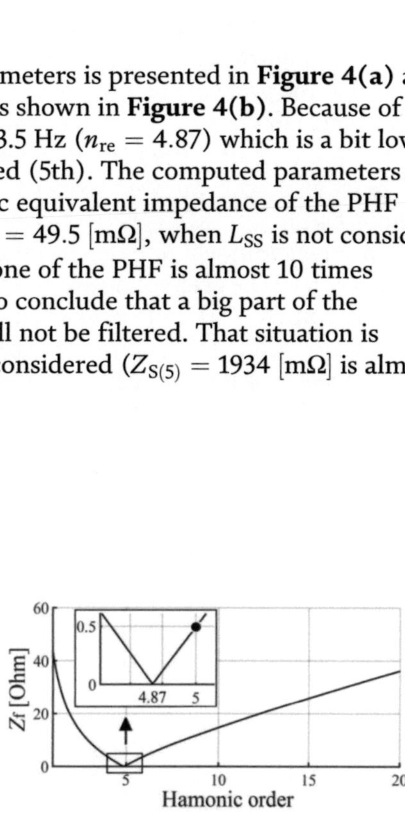

(b)

Figure 4.
(a) Expressions used to compute PHF parameters (q_{Lf}—Reactor quality factor, n_{re}—order of the resonance frequency), (b) PHF impedance versus frequency characteristic.

n_{re}	Q_f [Var]	U_f [V]	L_f [mH]	C_{fY} [μF]	q_{Lf}	$Z_{f(1)}$ [Ω]	$Z_{f(5)}$ [Ω]	L_{SS} [mH]	$Z_{S(5)}$ [Ω] (without L_{SS})	$Z_{S(5)}$ [Ω] (with L_{SS})
4.87	1210	230	6.1	69.73	150	43.71	0.49	1.2	0.0495	1.934

Table 1.
Computed equivalent parameters of the PHF (one-phase).

3.3 Simulation results

The simulation data before the PHF connection, with and without the grid side line reactor L_{SS} are compared based on: grid voltage and current waveforms, characteristic harmonics (**Figure 5**), THD, grid, and filter active and reactive powers (fundamental harmonic, **Figure 6**) as well as the electrical system impedance versus frequency characteristics seen from the thyristor bridge input (**Figure 7**).

Figure 5(a) represents the grid voltage waveforms and **Figure 5(b)** its spectrum (for the current—**Figure 5(c)** and **(d)**, respectively). It can be noticed that when the additional line reactor L_{SS} is connected (k1 open), all the harmonics amplitudes of the grid voltage have increased (**Figure 5(b)**), whereas the ones of the grid current have decreased (**Figure 5(d)**). The 5th harmonic current amplitude is therefore better reduced at the grid side (**Figure 5(d)**).

In **Figure 6(a)**, the grid active powers ($P_{S(1)}$) are almost the same and it can also be seen that after the PHF connection, in both cases (without and with L_{SS}) the grid reactive powers are well reduced but the reduction is a bit better in the case without L_{SS}.

With the additional line reactor, the PHF presents less power losses (**Figure 6(b)**); the grid voltage is more distorted (**Figure 6(c)**) and the grid current is less distorted (see THD in **Figure 6(c)**). The increase of the grid current total harmonic distortion (THD_{IS}—**Figure 6(c)**) after the PHF connection (case without L_{SS}) is due to fact that the grid current fundament harmonic amplitude has decreased because of the reactive power compensation and the harmonics have almost not been reduced (1).

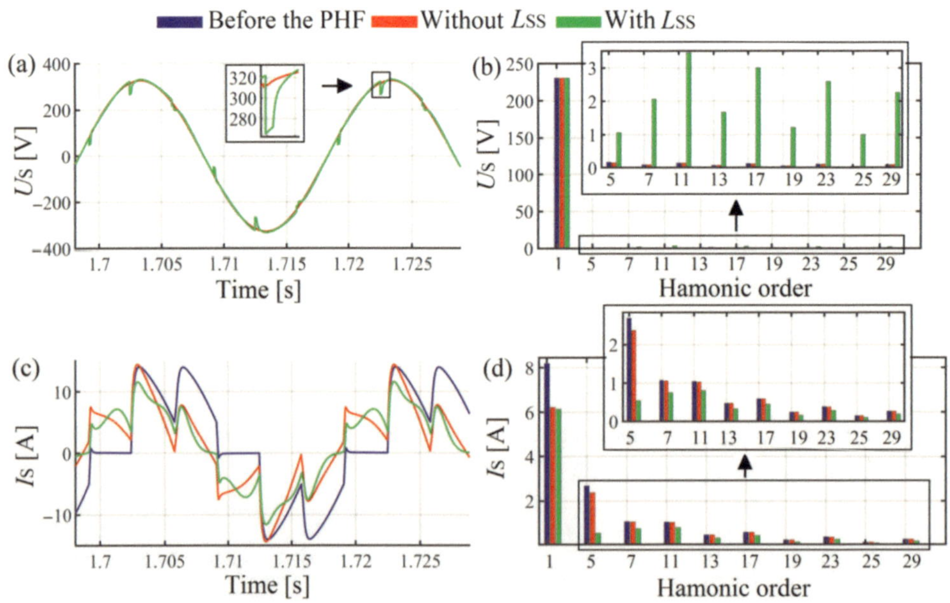

Figure 5.
(a) Grid voltage/current waveforms and their (b) spectrums (the red and green colors represent the cases when the filter is connected).

$$\text{THD}_\text{I} = \frac{\sqrt{\sum_{k=2}^{50} I_k^2}}{I_1} \tag{1}$$

Figure 6(d) presents the PHF work efficiency in terms of harmonics reduction. It represents the percentage of current harmonics flowing from the load to the grid after the PHF connection (without and with L_{SS}). It can be observed that without L_{SS} in the electrical system, almost 100% of current harmonics generated by the load flow to the grid. The connection of L_{SS} has improved the PHF work efficiency especially on the 5th harmonic reduction (**Figure 6(d)**).

The impedance versus frequency characteristics of the electrical system seen from the thyristor bridge input are presented in **Figure** 7. It can be observed in the both cases (without and with L_{SS}), the parallel and series resonances. The series resonance comes from the PHF and the parallel resonance comes from the parallel connection between the grid and the PHF. The resonances are below the 5th harmonic, which means that the filter was well deigned. Such harmonics (at the resonance) do not exist in considered system.

The performed studies have clearly showed that the PHF work efficiency in terms of harmonics mitigation strongly depends on the electrical grid parameters. Because of that it is necessary to obtain the information about the grid equivalent impedance for the harmonic to be eliminated (through the electrical system short-circuit power at the PCC) before the filter installation. In the case where that equivalent impedance is smaller than the one of the PHF, the solution can be the use of an additional line reactor between the PCC and the grid to increase the filtration effectiveness as it has been demonstrated in the chapter.

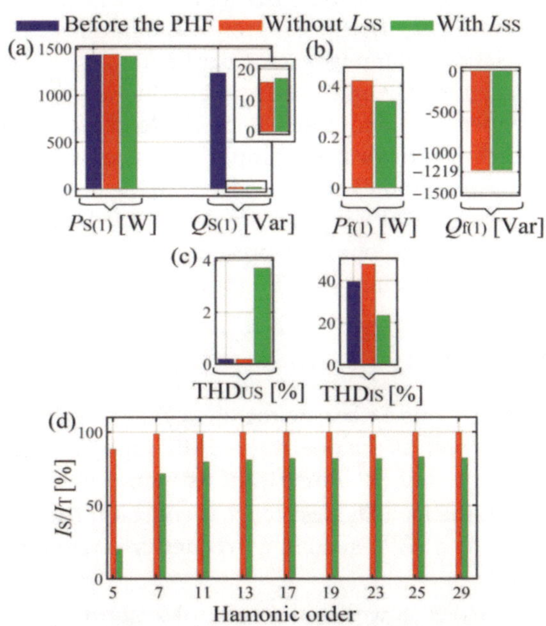

Figure 6.
(a) Grid active ($P_{S(1)}$) and reactive power ($Q_{S(1)}$), (b) filter active ($P_{f(1)}$) and reactive ($Q_{f(1)}$) power, (c) grid voltage and current THD, and (d) PHF work efficiency on harmonics reduction (the red and green colors represent the cases when the filter is connected).

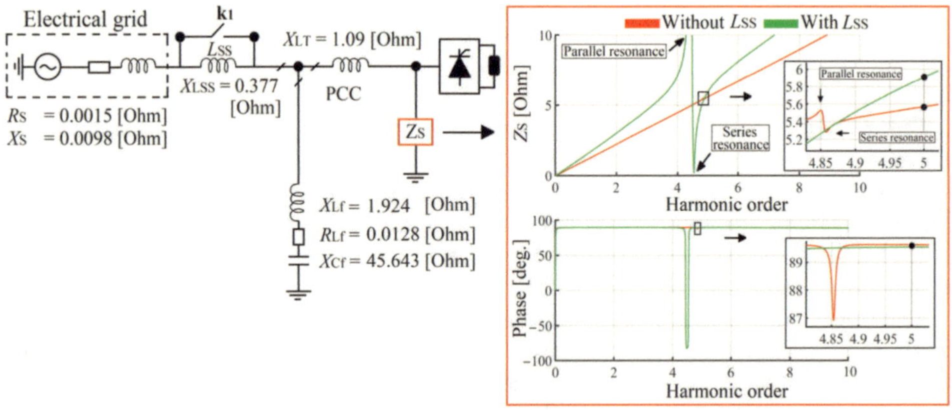

Figure 7.
Impedance versus frequency characteristics of the electrical system seen from the thyristor bridge input terminals.

4. Investigation on the influence of the thyristor bridge input reactor size on the SAPF work efficiency

The work efficiency of the SAPF does not depend only on the designed control system or inverter parameters, but also on the parameters of the electrical system to which it is connected. For instance, in the electrical system with a diode or thyristor bridge as load, the accepted operating efficiency of the SAPF (with input reactor as switching ripple filter) may not be met due to the high rate of load current change (di/dt) at the points of commutation notches. Therefore, it is necessary to study and know the electrical system (grid and load sides) before its installation. That problem is also mentioned in the literature [21]. In this chapter, it is demonstrated that to obtain a better SAPF work efficiency, its input reactor parameters (see $L_{\_inv}$ in **Figure 1**) should be selected based not only on the effective reduction of the inverter switching ripple or the control system demand, but also on the load parameters, such as the parameters of the diode or thyristor bridge input line reactor (see L_T in **Figure 1**). For a good clarification, two case studies are considered in the chapter: In the first one, it is presented the influence of the SAPF input reactor size (load input reactor constant) on its work efficiency. In the second one, it is presented the influence of the load input reactor on the SAPF work efficiency (SAPF input reactor constant).

4.1 Simulation assumptions

The SAPF control system and DC capacitor ($C_{\_inv}$) parameters are constant, the load input reactor resistance as well as the one at the SAPF input is neglected, the additional line reactor L_{SS} is not considered (k1 closed), the asymmetry resistance ($R_{asym} = 60\ \Omega$) is considered (k5 closed), the load DC resistance is constant, the switches k2 and k4 are opened, and k3 is closed. The SAPF switching frequency is fixed to 20 kHz.

4.2 Description of the SAPF as well as its control system

The SAPF presented in **Figure 1** is three legs three wires invert with an reactor ($L_{\_inv}$) at its input and capacitor ($C_{\_inv}$) at its DC side. Its proposed control system is presented in **Figure 8** and is organized in three control loops: (1)–(3).

The role of the *control loop (1)* is to maintain the SAPF DC voltage (U_{DC_inv}) constant and at the level as the given reference voltage (U_{DC_ref}) [32].

The *control loop (2)* algorithm is based on the time domain instantaneous *p-q* theory [33]. Its role is to split the distorted instantaneous load current (i_{Tabc}) into four different components related to (see **Figure 8**): active ($i_{abc\_(\overline{p})}$), reactive ($i_{abc\_(\overline{q})}$), asymmetry ($i_{abc\_(\mathrm{asym})}$), and harmonic ($i_{abc\_(harmo)}$). The output signal of the *control loop (2)* is therefore the instantaneous reference current ($i_{abc\_(\mathrm{ref})}$) which is constituted of three components (reactive, asymmetry, and harmonic currents) as shown in **Figure 8** (the active component ($i_{abc\_(\overline{p})}$) is not included).

Observing the *control loop (2)* algorithm, it can be noticed that the splitting of the instantaneous load current (i_{Tabc}) into different components starts from its measurement as well as the measurement of the instantaneous PCC voltage (u_{Sabc}). The supply voltage is filtered ("supply voltage filtration") to avoid its distortions to be found on the reference courant. The instantaneous real (p) and imaginary (q) powers are obtained after transforming the instantaneous PCC voltage and current from the *a-b-c* coordinates to the α-β rectangular coordinates. The low-pass filter (LPF) and band-pass filter (BPF) are used to filter the instantaneous real (p) and imaginary (q) powers so that their components (($\overline{p}, \overline{q}$)—constant components related to the fundamental harmonic (positive sequence), ($\tilde{p}_{\mathrm{h}}, \tilde{q}_{\mathrm{h}}$)—component related to the current harmonic and ($\tilde{p}_{2\mathrm{n}}, \tilde{q}_{2\mathrm{n}}$)—component related to the current asymmetry (negative sequence of the fundamental harmonic)) can be used in the matrix (see **Figure 8**) to compute the instantaneous currents (i_α, i_β) in α-β axes. After the matrix computation, different current components are obtained: ($i_{\alpha(\overline{\mathrm{p}})}, i_{\beta(\overline{\mathrm{p}})}$)—instantaneous real current (fundamental harmonic) in α and β axes respectively, (($i_{\alpha(\tilde{\mathrm{p}}_{2\mathrm{n}})}, i_{\alpha(\tilde{\mathrm{q}}_{2\mathrm{n}})}$), ($i_{\beta(\tilde{\mathrm{p}}_{2\mathrm{n}})}, i_{\beta(\tilde{\mathrm{q}}_{2\mathrm{n}})}$))—instantaneous asymmetry current in α and β axes, respectively, (($i_{\alpha(\tilde{\mathrm{p}}_{\mathrm{h}})}, i_{\alpha(\tilde{\mathrm{q}}_{\mathrm{h}})}$), ($i_{\beta(\tilde{\mathrm{p}}_{\mathrm{h}})}, i_{\beta(\tilde{\mathrm{q}}_{\mathrm{h}})}$))—instantaneous harmonic current in α and β axes respectively, ($i_{\alpha(\overline{\mathrm{q}})}, i_{\beta(\overline{\mathrm{q}})}$)—instantaneous imaginary current (fundamental harmonic) in α and β axes respectively. The components such as $i_{\alpha(2\mathrm{n})}$ and $i_{\beta(2\mathrm{n})}$ are the instantaneous asymmetry current in α and β axis respectively and the components such as $i_{\alpha(\mathrm{h})}$ and $i_{\beta(\mathrm{h})}$ are the instantaneous harmonics current in α and β axis respectively.

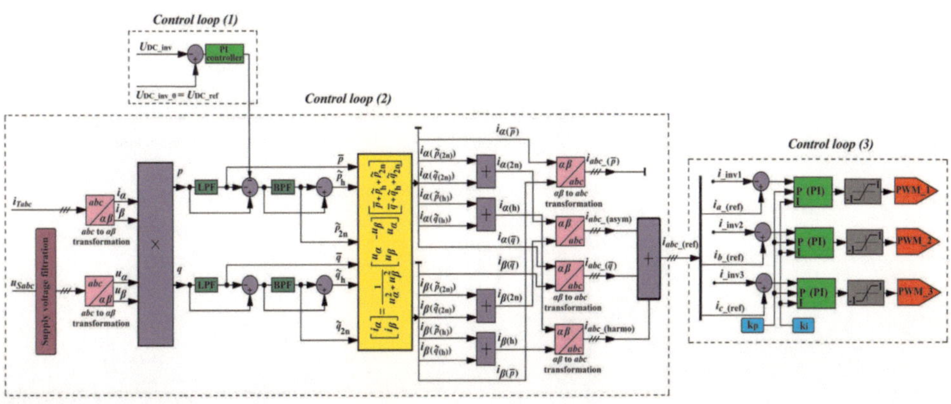

Figure 8.
Block diagram of the SAPF control system: a, b, and c—*Phase designations of the supply network; 1, 2, and 3—represent the inverter inputs.*

ΔU_{L_inv} [V]	$L_{\_inv\_min}$ [mH]	$L_{\_inv\_max}$ [mH]	$I_{a\_(ref)}$ [A]	$I_{b\_(ref)}$ [A]	$I_{c\_(ref)}$ [A]	k_{DC}	U_{S_p-p} [V]	$U_{DC\_inv\_0}$ [V]	ΔW_{DC_inv} [J]	ΔU_{DC_inv} [V]	$C_{\_inv}$ [mF]
3	1.4	—	4.46	9.38	7.34	1.32	400	750	11	5	3
16	—	7.2									

Table 2.
Computed SAPF parameters.

In the *control loop (3)*, the instantaneous reference current ($i_{abc\_(ref)}$) is compared to the feedback loop current coming from the inverter input ($i_{\_inv123}$). It can also be seen the blocks of PI controller, saturation, and pulse-width modulation (PWM) system.

4.2.1 Computation of the SAPF parameters

The proposed expressions used to compute the SAPF parameters are presented in (2)-(4) and the computed parameters in **Table 2**. The PI controller parameters are presented in **Table 3**.

$$U_{DC\_inv\_0} > \mathbf{k}_{DC}\left(\sqrt{2}\, U_{S\_p-p}\right), \mathbf{k}_{DC} > 1 \tag{2}$$

$$L_{\_inv} = \frac{3\Delta U_{L\_inv}}{\omega_{(1)}\left(I_{a\_(ref)} + I_{b\_(ref)} + I_{c\_(ref)}\right)} \tag{3}$$

$$C_{\_inv} = \frac{\Delta W_{DC\_inv}}{U_{DC\_inv\_0}\Delta U_{DC\_inv}} \tag{4}$$

Where: U_{S_p-p}—phase-to-phase PCC voltage, $\mathbf{k}_{DC}$—coefficient, ΔU_{L_inv}—inverter input reactor voltage drops, $I_{a\_(ref)}$, $I_{b\_(ref)}$, and $I_{c\_(ref)}$—the reference currents from the output of the *control loop (2)*, ΔW_{DC_inv}—DC capacitor energy variation between the max and the min, ΔU_{DC_inv}—DC capacitor voltage variation between the max and the min.

4.3 Simulation results

The grid voltage and current waveforms with spectrums before the SAPF connection (L_T = 0.025 mH) are constituted in **Figures 9** and **10** respectively. The current waveforms asymmetry can be observed in **Figure 10** (R_{asym} = 60 Ω).

	kp	ki
Control loop (1)	40,000	43.75
Control loop (3)	250	0.0001

Table 3.
PI controller parameters.

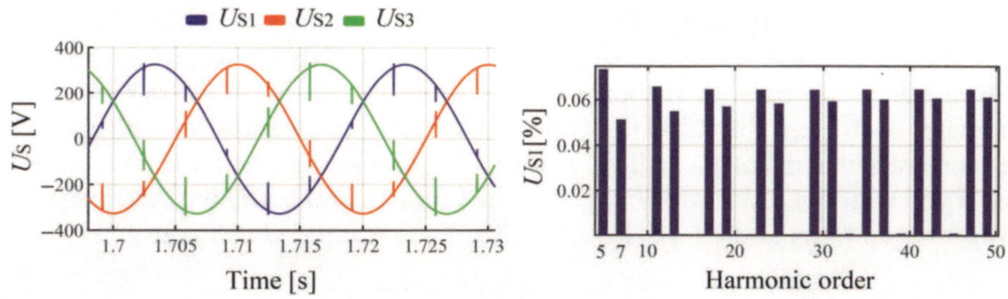

Figure 9.
Waveforms of PCC voltage and spectrum before the SAPF connection.

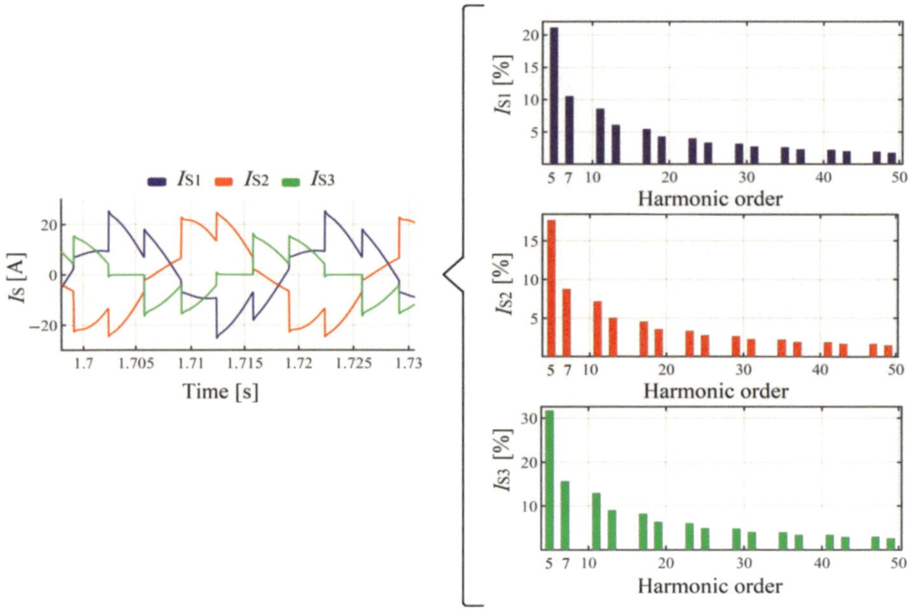

Figure 10.
Waveforms of the grid current with the spectrums before the SAPF connection.

4.3.1 Influence of the inverter input reactor size on the SAPF work efficiency

The goal of the investigation is to show how the size of the inverter input reactor has an influence on the SAPF work efficiency. For this purpose, two values of the inverter input reactor inductance are considered (**Table 2**): The first one ($L_{\_inv\_min} = 1.4$ mH) is obtained after assuming a minimum reactor voltage drop (e.g., $\Delta U_{L\_inv\_min} = 3$ V) and the second one ($L_{\_inv\_max} = 7.2$ mH) after assuming a maximum reactor voltage drop (e.g., $\Delta U_{L\_inv\_min} = 16$ V).

The grid voltage and current waveforms as well as the ones of the SAPF are presented respectively in **Figures 11** and **12**.

On the one hand, the SAPF with minimum input reactor inductance $L_{\_inv\_min}$ (see **Figures 11(a)** and **12(a)**) presents higher switching ripple components than the SAPF with the maximum input reactor inductance $L_{\_inv\_max}$ (see **Figures 11(b)** and **12(b)**). On the other hand, it presents better shape of the grid current waveforms at the commutation notches (comparing the grid current waveforms in **Figure 11(a)** to those in **Figure 11(b)**).

The grid voltage and current THD, powers, and asymmetry coefficient (k_{asym}) before the SAPF connection are presented in **Table 4**. In **Table 5**, it can be seen that the SAPF with $L_{\_inv\_min}$ presents the lowest grid current and voltage THD as well as asymmetry coefficient. In **Table 6**, the SAPF with $L_{\_inv\_max}$ has the lowest grid voltage true total harmonic distortion (TTHD) and the highest grid current TTHD (see (5)) in comparison with the SAPF with $L_{\_inv\_min}$. The fundamental harmonic active, reactive, and apparent powers at the PCC, load, and SAPF are compared in **Table 7**.

$$\mathrm{TTHD_{Is}} = \frac{\sqrt{I^2_{\mathrm{S\_true\_RMS}} - I^2_{\mathrm{S(1)}}}}{I_{\mathrm{S(1)}}} \tag{5}$$

DOI: http://dx.doi.org/10.5772/intechopen.1002734

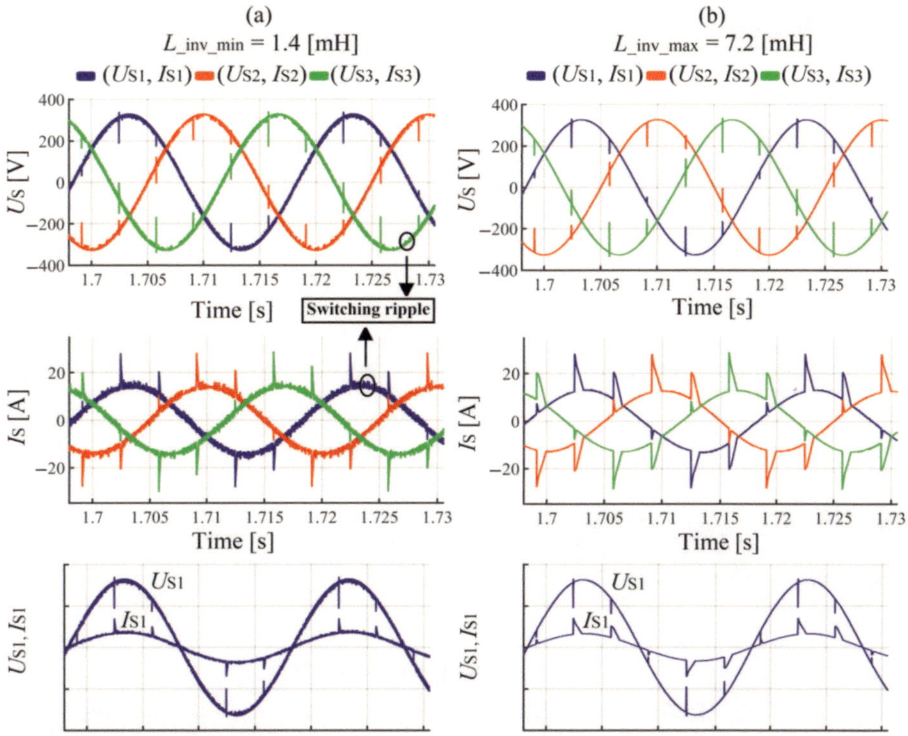

Figure 11.
Waveforms of the grid voltage and current after the SAPF connection: (a) for $L_{\_inv\_min}$ and (b) for $L_{\_inv\_max}$.

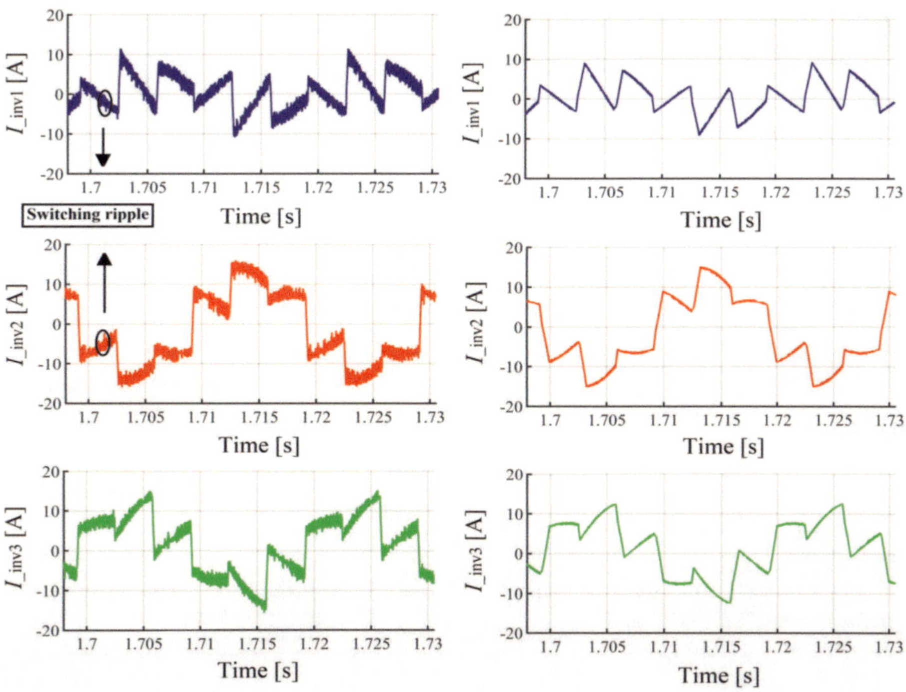

Figure 12.
Waveforms of the SAPF current: (a) for $L_{\_inv\_min}$ and (b) for $L_{\_inv\_max}$.

	THD_{US} [%]	THD_{IS} [%]	$Q_{S(1)}$ [Var]	$P_{S(1)}$ [W]	$S_{S(1)}$ [VA]	k_{asym} [%]
L1	0.25	28.07	414.08	2838.4	2868.5	33.25
L2	0.25	23.41	1954.5	2833.1	3441.9	
L3	0.25	42.07	1181.8	1507.2	1915.2	

Table 4.
Grid voltage and current parameters before the SAPF connection.

	$L_{\_inv\_min}$ = 1.4 [mH]			$L_{\_inv\_max}$ = 7.2 [mH]		
	THD_{US} [%]	THD_{IS} [%]	k_{asym} [%]	THD_{US} [%]	THD_{IS} [%]	k_{asym} [%]
L1	0.18	10.34	0.59	0.25	26.7	0.75
L2	0.19	10.99		0.25	27.92	
L3	0.18	10.93		0.25	27.7	

Table 5.
Grid voltage and current parameters after the SAPF connection.

	After the SAPF connection			
	$L_{\_inv\_min}$ = 1.4 [mH]		$L_{\_inv\_max}$ = 7.2 [mH]	
	$TTHD_{US}$ [%]	$TTHD_{IS}$ [%]	$TTHD_{US}$ [%]	$TTHD_{IS}$ [%]
L1	3.57	16.31	1.70	27.09
L2	2.93	16.19	1.50	27.59
L3	3.60	16.20	2.13	27.89

Table 6.
PCC voltage and grid current TTHD.

		$L_{\_inv\_min}$ = 1.4 [mH]			$L_{\_inv\_max}$ = 7.2 [mH]		
		PCC	Load	SAPF	PCC	Load	SAPF
$P_{S(1)}$ [W]	L1	2411.4	2838.8	427.40	2389.3	2838.8	449.82
	L2	2535.8	2837.9	438.19	2416.8	2837.9	421.33
	L3	2522.3	1505.7	−890.35	2397.5	1505.7	−891.61
$Q_{S(1)}$ [Var]	L1	46.30	414.13	370.22	91.78	414.13	323.23
	L2	57.54	1950.4	1898	97.07	1950.4	1854.5
	L3	39.62	1180.6	1143.7	117.25	1180.6	1062.6
$S_{S(1)}$ [VA]	L1	2411.9	2868.8	565.46	2391.1	2868.8	553.92
	L2	2536.5	3443.5	1948	2418.8	3443.5	1.901.8
	L3	2522.6	1913.3	1449.4	2400.3	1913.3	1387.1

Table 7.
Active, reactive, and apparent powers at the PCC, load, and SAPF for the minimum and maximum input inverter reactor inductance.

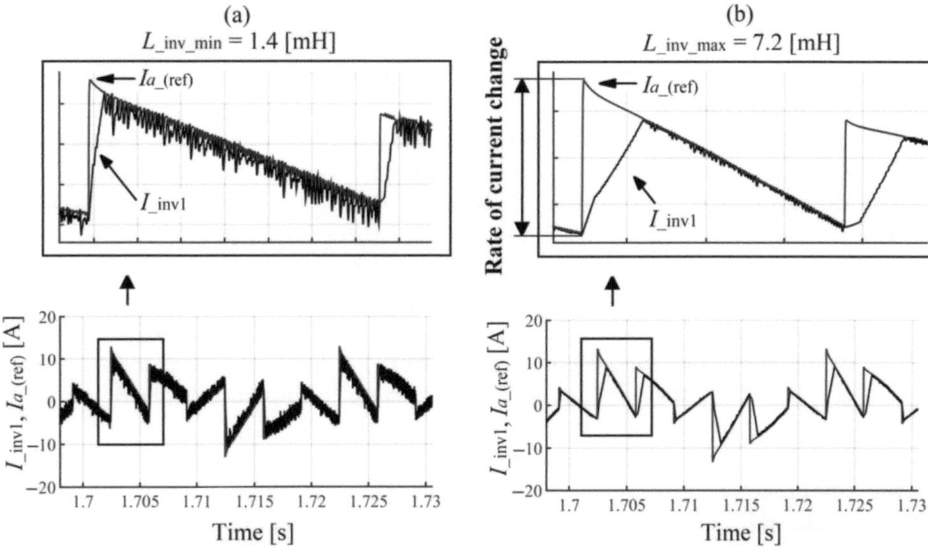

Figure 13.
Example of waveforms comparison between the reference and compensating currents (see control loop (3) in ***Figure 8****): (a) for* $L_{\_inv\_min}$ *and (b) for* $L_{\_inv\_max}$.

An example of comparison waveforms between the reference current ($I_{a\_(ref)}$) and the compensating current (I_{inv1}) is presented in **Figure 13(a)** and **(b)**. In the case of the SAPF with $L_{\_inv\_min}$ (**Figure 13(a)**) as well as in the case of the SAPF with $L_{\_inv\_max}$ (**Figure 13(b)**), the compensating current (I_{inv1}) has difficulty to track the reference current ($I_{a\_(ref)}$) at the points of commutation notches, due to the high rate of the reference current change at those points. That tracking difficulty is more accentuated in the case of the SAPF with $L_{\_inv\_max}$ (**Figure 13(b)**) than in the case of the SAPF with $L_{\_inv\_min}$ (**Figure 13(a)**). The high rate of current change observed on the reference current waveforms comes from the load current (I_T). Because of the small size of the thyristor bridge input reactor L_T, the rate of current change (di/dt) during commutation is high.

In this case study, it can be noticed that the gap between the reference current and the compensating current observed in **Figure 13(a)** and **(b)** is one of the factors that can influence the SAPF work efficiency, mostly in terms of current harmonics mitigation (see grid current THD in **Table 5**). That gap is responsible for the high amplitude ripples at the points of commutation notches observed on the grid current waveform in **Figure 11(b)**.

It has been clearly demonstrated that the size of the SAPF input reactor affects its work efficiency. The SAPF with big size of input reactor has showed a better result in terms of inverter switching ripples reduction but worst result in terms of reducing the ripples caused by the high rate of current change during the commutation. In contrary to the SAPF with big size reactor, the SAPF with small size of input reactor has showed a worst result in terms of inverter switching ripples reduction but a better result in terms of ripple reduction at the high rate of current change during the commutation. At the grid side, the SAPF with small inductance ($\boldsymbol{L_{\_inv\_min}}$) has showed better results in terms of THD (TTHD current), asymmetry coefficient, and reactive power compensation (**Table 7**) at the grid side. The SAPF with big reactor inductance ($\boldsymbol{L_{\_inv\_max}}$) has showed better results in terms of TTHD of the PCC voltage.

4.3.2 Influence of the thyristor bridge input reactor size on the SAPF work efficiency

This investigation is performed because of the problem of ripples reduction due to the high rate of load current change observed in the previous investigations. The goal is to demonstrate that the SAPF input reactor size should be chosen by considering also the thyristor bridge input reactor size in order to improve the SAPF work efficiency.

The SAPF parameters are constant during the investigation. The value of the SAPF input reactor inductance ($L_{\_inv\_min} = 1.4$ mH) is chosen based on the previous investigation. The only load parameter changing is the input reactor L_T (0.25, 1.4, or 3 mH). Three case studies are compared: (a) The SAPF input reactor inductance is higher than the thyristor-bridge input reactor inductance ($L_{\_inv\_min} > L_T$), (b) the SAPF input reactor inductance is equal to the thyristor-bridge input reactor inductance ($L_{\_inv\_min} = L_T$), and (c) the SAPF input reactor inductance is smaller than the thyristor-bridge input reactor inductance ($L_{\_inv\_min} < L_T$).

The grid voltage and current waveforms in **Figure 14** show that in the cases of SAPF with the input reactor inductance equal or smaller than thyristor bridge input reactor, the ripples caused by the high rate of current change at the points of commutation notches are better reduced (see current waveform in **Figure 14(b)** and **(c)**). The case in which $L_{\_inv\_min} < L_T$ presents the best shape of grid current waveforms (**Figure 14(c)**).

In **Table 8**, it can be observed that the best results in terms of grid voltage and current THD as well as asymmetry mitigation are when $L_{\_inv\_min} \le L_T$. Observing the grid powers in **Table 9**, it can be seen that the case with $L_{\_inv\_min} < L_T$ has the best reactive power compensation.

The TTHD of the grid voltage and current are presented in **Table 10**. The case with $L_{\_inv\_min} < L_T$ presents the best result in terms of current $TTHD_{IS}$ and the worst results in terms of voltage $TTHD_{US}$.

The waveforms of the reference current compared to the ones of the compensating current are presented in **Figure 15**. It can be seen that in the case of $L_{\_inv\_min} < L_T$, the reference current and the compensating current match each other (**Figure 15(c)**). The increase of the thyristor-bridge input reactor inductance L_T to a value equal or higher than the SAPF input reactor inductance ($L_{\_inv\_min}$) has reduced the rate of current change at the points of commutation notches making possible the tracking of the reference current by compensating current (**Figure 15(b)** and **(c)**).

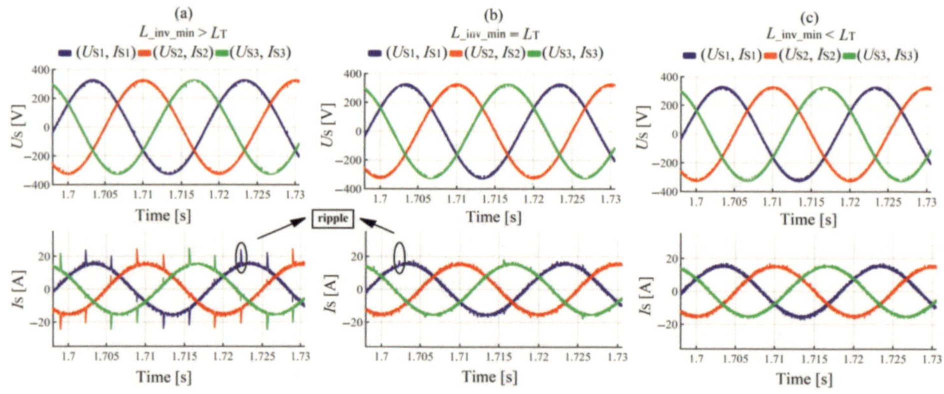

Figure 14.
Waveforms of PCC voltage and current for different value of the thyristor bridge input reactor: (a) $L_{\_inv\_min} > L_T$, (b) $L_{\_inv\_min} = L_T$, (c) $L_{\_inv\_min} < L_T$.

	Before the SAPF connection			**After the SAPF connection**		
	L_T = 1 nH			**$L_{\_inv\_min} > L_T$**		
	THD_{US} [%]	**THD_{IS} [%]**	**k_{asym} [%]**	**THD_{US} [%]**	**THD_{IS} [%]**	**k_{asym} [%]**
L1	0.25	28.07	33.25	0.15	8.77	0.30
L2	0.25	23.41		0.15	8.75	
L3	0.25	42.07		0.15	8.77	
	After the SAPF connection					
	$L_{\_inv\_min} = L_T$			$L_{\_inv\_min} < L_T$		
	THD_{US} [%]	THD_{IS} [%]	k_{asym} [%]	THD_{US} [%]	THD_{IS} [%]	k_{asym} [%]
L1	0.04	2.47	0.30	0.02	1.22	0.20
L2	0.05	2.91		0.02	1.49	
L3	0.05	2.79		0.02	1.22	

Table 8.
Grid voltage and current THD as well as asymmetry coefficient before and after the SAPF connection.

	Before the SAPF connection		**After the SAPF connection**	
	L_T = 1 nH		**$L_{\_inv\_min} > L_T$**	
	$Q_{S(1)}$ [Var]	**$P_{S(1)}$ [W]**	**$Q_{S(1)}$ [Var]**	**$P_{S(1)}$ [W]**
L1	414.08	2838.4	25.15	2402.5
L2	1954.5	2833.1	33.51	2400.1
L3	1181.8	1507.2	29.24	2393.2
	After the SAPF connection			
	$L_{\_inv\_min} = L_T$		$L_{\_inv\_min} < L_T$	
	$Q_{S(1)}$ [Var]	$P_{S(1)}$ [W]	$Q_{S(1)}$ [Var]	$P_{S(1)}$ [W]
L1	24.86	2374.8	20.42	2340.2
L2	29.04	2377.1	28.59	2340.1
L3	28.98	2372.4	24.43	2333.3

Table 9.
Grid voltage and current reactive and active power before and after the SAPF connection.

	After the SAPF connection					
	$L_{\_inv\_min} > L_T$		**$L_{\_inv\_min} = L_T$**		**$L_{\_inv\_min} < L_T$**	
	$TTHD_{US}$ [%]	**$TTHD_{IS}$ [%]**	**$TTHD_{US}$ [%]**	**$TTHD_{IS}$ [%]**	**$TTHD_{US}$ [%]**	**$TTHD_{IS}$ [%]**
L1	3.12	12.60	3.18	7.16	3.30	6.59
L2	2.77	13.77	2.85	8.08	2.92	6.82
L3	3.12	13.40	3.23	7.20	3.29	6.11

Table 10.
PCC voltage and grid current TTHD for different value of thyristor bridge input reactor inductance.

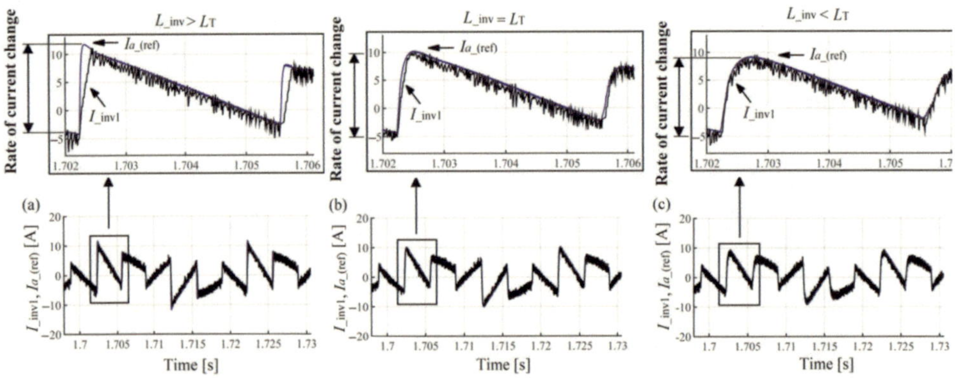

Figure 15.
Waveforms comparison between the reference and compensating current: (a) $L_{\_inv\_min} > L_T$, (b) $L_{\_inv\_min} = L_T$, (c) $L_{\_inv\_min} < L_T$.

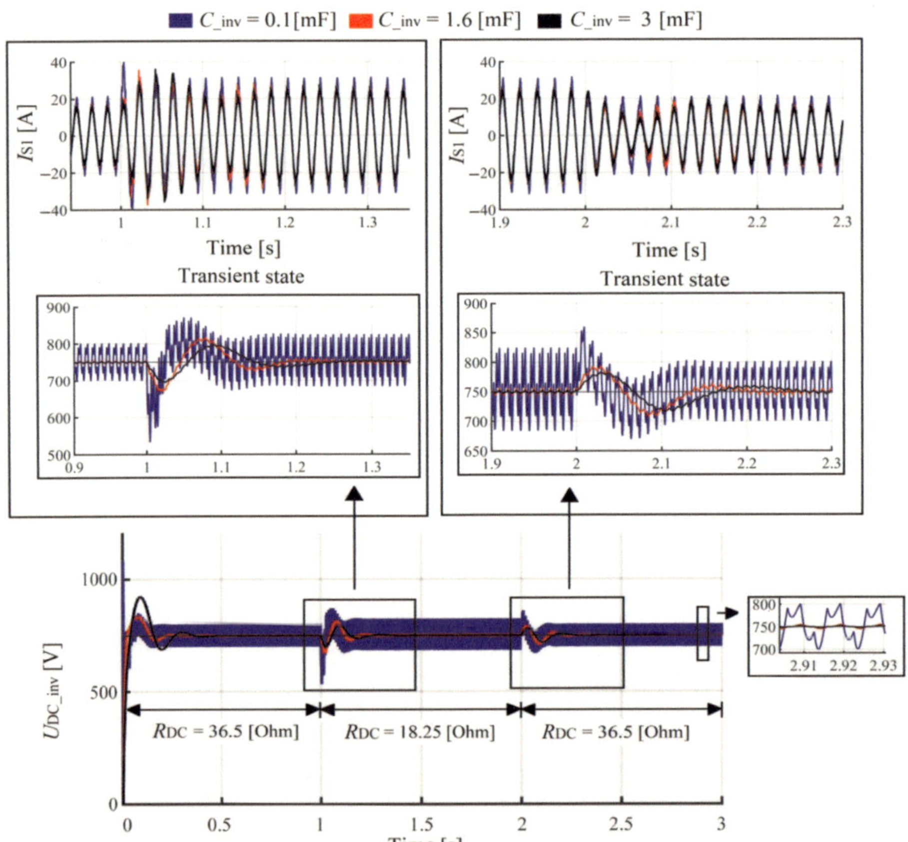

Figure 16.
Capacitor DC voltage waveforms for different capacitance: Transient state observation after the thyristor-bridge DC resistance change (R_{DC}).

Figure 16 presents the SAPF DC capacitor voltage for different values of capacity. It can be also observed the ability of the control system algorithm in response to the decrease and increase of the thyristor-bridge DC resistance (load change).

The investigations have clearly showed that the size of the thyristor-bridge input reactor has an influence of the SAPF work efficiency. The SAPF with the input reactor inductance equal or smaller than the one at the thyristor-bridge input has presented the best results in terms of grid voltage and current THD as well as in terms of reactive power and asymmetry mitigation. The inductance of the SAPF input reactor should be computed or chosen considering the size of the thyristor-bridge input reactor (see expression (6)). ΔU_{L_inv} should be chosen in such a way to obtain $L_{\_inv} \leq L_T$.

$$L_{\_inv} = \frac{3\Delta U_{L\_inv}}{\omega_{(1)}\left(I_{a\_(ref)} + I_{b\_(ref)} + I_{c\_(ref)}\right)} \leq L_T \tag{6}$$

5. Investigation on the work efficiency of the SHAPF

In the chapter, the investigation of the SHAPF work efficient is focused on the tuning frequency of the passive part (balance load). The passive part is tuned to three different frequencies and the results in terms of grid voltage and current THD, waveforms and powers are compared. The influence of the thyristor bridge input reactor on the choice of the passive part tuning frequency is presented as well as the limitation of such topology in terms of asymmetry compensation. It is also demonstrated that the grid dependency of the passive part work efficiency in terms of harmonics mitigation (see investigation in Chapter 3) is eliminated as well as the parallel resonance between the passive part and the grid.

5.1 Functionality principle of the SHAPF

The SHAPF work principle is presented in **Figure 17**. The load is symmetrical and is the source of distorted current, which is composed of three components: active ($I_{T(1)\_Active}$), inductive reactive ($I_{T(1)\_Inductive\_Reactive}$), and harmonics ($I_{T(h)}$). The SHAPF passive part is tuned to the frequency of 7th harmonic just to illustrate the SHAPF functionality principle.

The role of the passive part is to compensate the load fundamental harmonic reactive power by producing through its capacitor a capacitive reactive current ($I_{f(1)\_Capacitive\_Reactive}$) which at the PCC cancel with the load inductive reactive current ($I_{T(1)\_Inductive\_Reactive}$). In **Figure 17**, it can be seen that, the load inductive reactive current which was 5.45 A is reduced to 0.22 A at the grid side. It plays also the role of the inverter switching ripples attenuation mostly through its reactor. It has small impedance for the resonance frequency and frequencies around.

The SHAPF proposed control system is designed without any feedback signal coming from the inverter AC side (see **Figures 17** and **18**). As in the case of the SAPF control system (see **Figure 8**), it contains also three control loops ((1) to (3)). Its algorithm (based on the instantaneous *p-q* theory) computes the reference current ($I_{f(h)}$) using the grid current remaining harmonics ($I_{S(h)Remaining}$). Therefore based on the reference harmonics current $I_{ref(h)}$ (e.g., $I_{ref(5)} = 0.14e^{j42.5^{\circ}}$), the inverter generates current harmonics $I_{f(h)}$ (e.g., $I_{f(5)} = 2.62e^{-j48.9^{\circ}}$) with the same amplitudes and angle as the load current harmonics (e.g., $I_{T(5)} = 2.63e^{j128^{\circ}}$) but in opposite sign so that they cancel each other at the PCC (the grid current THD is reduced from 37.16 to 3.96%—**Figure 17**). The inverter behaves like a harmonics voltage source (see $U_{inv(h)}$ in

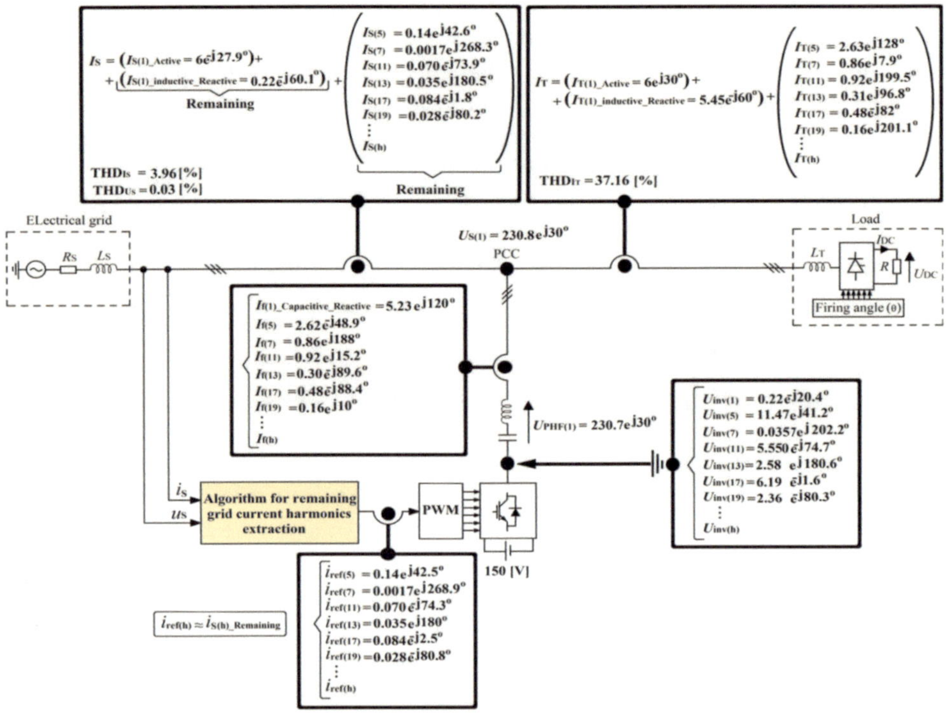

Figure 17.
SHAPF work principle.

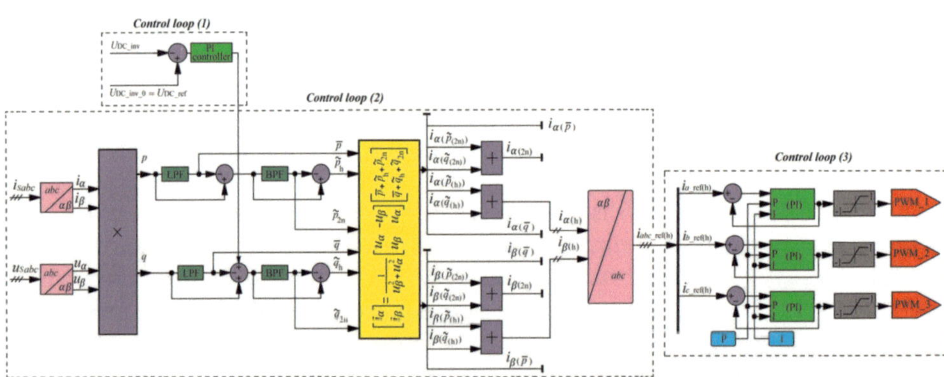

Figure 18.
Block diagram of the SHAPF control system: a, b, and c—*Phase designations of the supply network; 1, 2, and 3—represent the inverter inputs.*

Figure 17). At the grid side, it can be noticed a current (I_S) with almost not-change in the active component ($I_{S(1)\_Active}$) and with the remaining reactive and harmonic components (see **Figure 17**).

5.2 HAPF simulation studies

Three simulation case studies are investigated and compared: In the first one, the SHAPF passive part is tuned to the frequency lower (e.g., 4.87) than the frequency of the first load characteristic harmonic (the 5th); in the second one, it is tuned to the frequency lower (e.g., 6.87) than the frequency of the second load characteristic harmonic (the 7th), and in the third one, it is tuned to the frequency lower (e.g., 10.87) than the frequency of the third load characteristic harmonic (the 11th).

5.2.1 Simulation assumption

In the electrical system (see **Figure 1**), k1 is closed (L_{SS} is not considered), the load (k5 open) and the SHAPF control system parameters are constant, the load input reactor resistance as well as the ones of the passive part capacitor and inverter DC capacitor are neglected, k2 and k3 are opened, and k4 is close. The DC voltage of the active part is 150 V. The only parameter changing is the tuning frequency of the filter passive part. Because of the symmetrical power system, some results are presented only for one-phase.

5.2.2 SHAPF parameters computation

The SHAPF and the load ($Q_f = Q_T$, L_T) parameters are shown in **Table 11**. The passive part parameters are computed based on the formula presented in **Figure 4(a)**. Observing **Table 11**, it can be seen that, with the resonance frequency increase from 4.87 to 10.87, the passive part reactor inductance has decreased, whereas its capacitor capacity has increased. From the cost and power losses point of view, the PHF with the resonance frequency ($n_{re} = 10.87$ (543.5 Hz)) near the frequency of the 11th harmonic presents less exploitation cost and will generate less power losses because of its lowest equivalent resistance value.

Table 12 presents the electrical grid and SHAPF passive part equivalent impedances of the 5th, 7th, and 11th harmonics. The grid equivalent impedances for the 5th and 7th harmonics are smaller than the ones of the SHAPF passive part and concerning the 11th harmonic, the situation is contrary (see **Table 12**). The passive part impedance versus frequency characteristics are presented in **Figure 19**.

n_{re}	L_f [mH]	C_f [μF]	R_{Lf} [mΩ]	Q_f [Var]	q_{Lf}	L_T [mH]	U_{DC_inv} [V]
4.87	6.1	69.74	12.8	1210	150	3.5	150
6.87	3.0	71.27	6.3				
10.87	1.2	72.19	2.5				

Table 11.
SHAPF parameters.

$Z_{S(5)}$ [mΩ]	$Z_{SHAPF(5)}$ [mΩ]	$Z_{S(7)}$ [mΩ]	$Z_{SHAPF(7)}$ [mΩ]	$Z_{S(11)}$ [mΩ]	$Z_{SHAPF(11)}$ [mΩ]
49.5	490	69.2	243	108.8	96.4

Table 12.
Electrical grid and SHAPF passive part equivalent impedance of the 5th, 7th, and 11th harmonics (the line reactor L_{SS} is not considered).

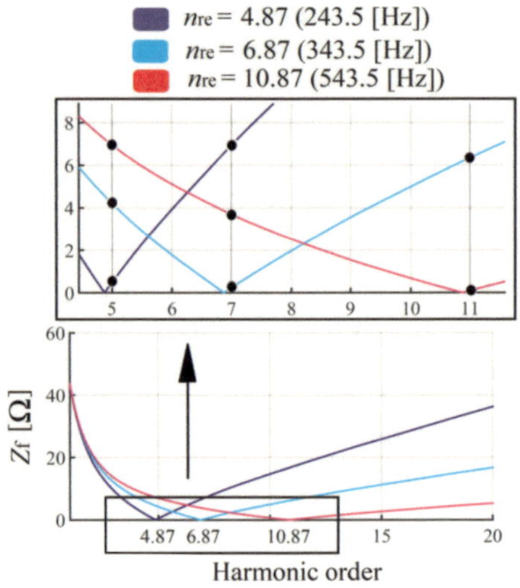

Figure 19.
SHAPF passive part impedance versus frequency characteristics.

5.2.3 Simulation results

The grid voltage and current waveforms with the spectrums before the SHAPF connection are presented in **Figure 20**. Because of the electrical grid rigidity (very small inductance), the PCC voltage waveforms are almost not distorted.

In **Figure 21**, the waveforms of the grid voltage (U_S) and current (I_S) as well as SHAPF current (I_f) after the SHAPF connection are presented. It can be noticed that the SHAPF with the passive part tuned to the frequency (e.g., 543.5 Hz, 10.87) a bit lower than the frequency of the 11th harmonic presents the best reduction of the grid current ripples at the high rate of current change points (see commutation points on

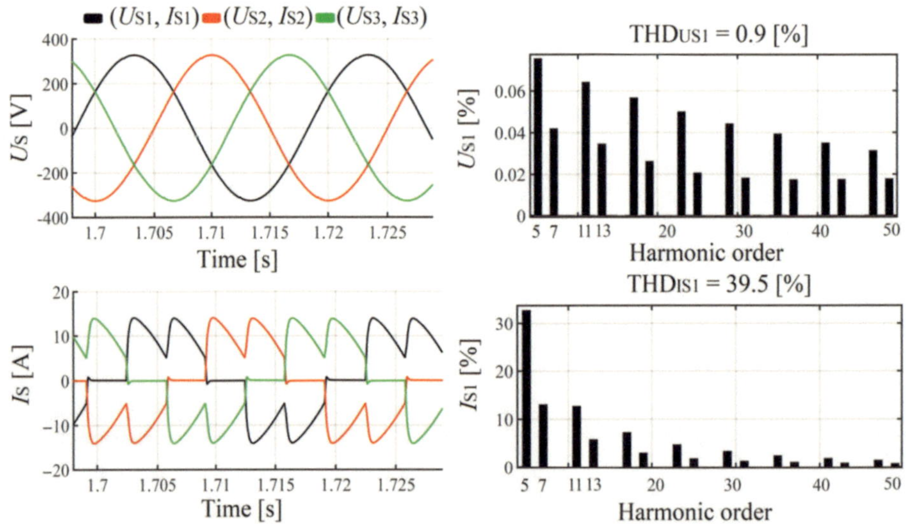

Figure 20.
Grid voltage (U_S) and current (I_S) waveforms with their spectrums before the SHAPF connection.

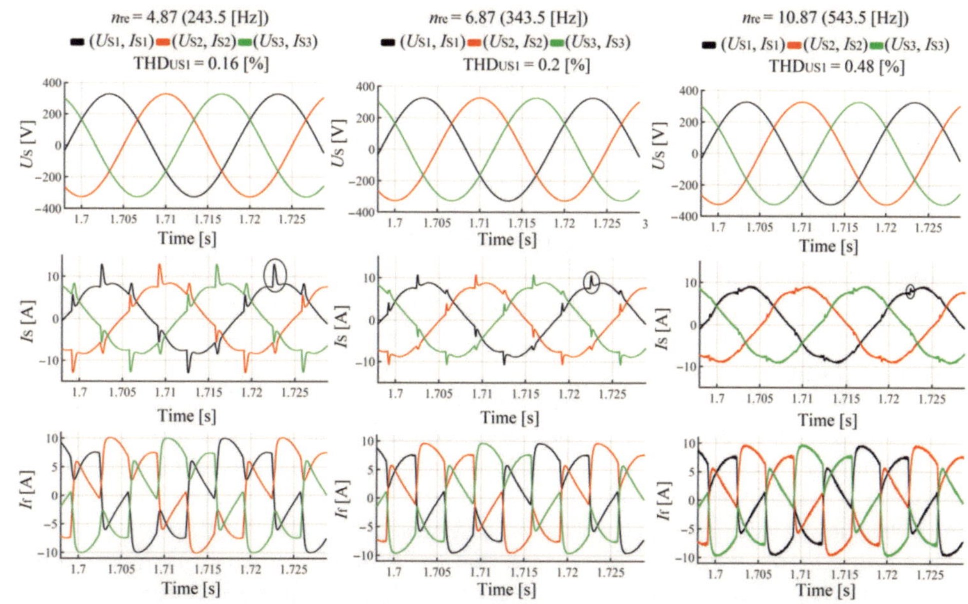

Figure 21.
Waveforms of grid voltage (U_S) and current (I_S) and SHAPF current (I_f) for the PHF tuned to the harmonic component frequency of order 4.87, 6.87, and 10.87.

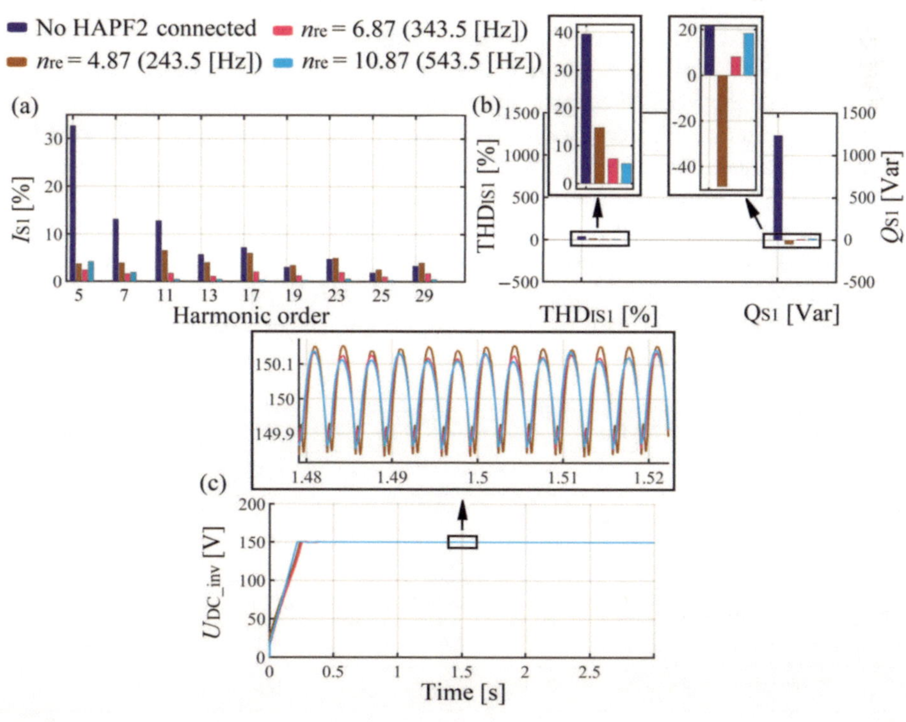

Figure 22.
(a) Grid current spectrum, (b) grid current THD and fundamental harmonic reactive power, and (c) SHAPF DC voltage.

the grid current waveforms (I_S) in **Figure 21**), although it presents the highest PCC voltage THD (**Figure 21**).

In the comparison spectrum of the grid current presented in **Figure 22(a)**, the SHAPF with the PHF tuned to the harmonic component frequency of 10,87 presents the best reduction of the higher harmonics starting from the 11th (according to the characteristics in **Figure 19**, the PHF presents the lowest impedance for the higher harmonics) as well as the lowest THD_{IS} (see **Figure 19(b)**). But it presents the highest 5th harmonic impedance (see **Figure 19**) and for that reason the 5th harmonic is worst reduced at the grid side (**Figure 22(a)**).

The SHAPF with the PHF tuning frequency ($n_{re} = 6.87$) around the frequency of the 7th harmonic presents the lowest amplitude of the grid current 5th and 7th harmonics (**Figure 22(a)**. Despite the fact that the grid equivalent impedances of the 5th and 7th harmonics are smaller than the ones of the SHAPF passive part (See **Table 12**), those harmonics are considerably reduced at the grid side after the SHAPF connection (see spectrum in **Figure 22(a)**) and this shows that the grid dependency of the PHF work efficiency is eliminated. The inverter through its control system has improved the passive part work efficiency in terms of harmonics filtration.

In **Figure 22(b)**, it can also be noticed that the PCC fundamental harmonic reactive power (Q_S) is the best compensated for the SHAPF with the PHF tuned to the harmonic component frequency of 6.87 (near the 7th).

The difference observed in the results of the grid fundamental reactive powers after the SHAPF connection in **Figure 22(b)** is due to the fact that, for different PHF tuning frequency, the SHAPF has generated different fundamental harmonic reactive power (a bit different than the one used to compute the PHF parameters in **Table 11**) for the compensation at the PCC (see $Q_{f(1)}$ in **Table 13**).

In comparison with the SAPF DC voltage (which is around 750 V in **Figure 16**), the SHAPF inverter DC voltage in **Figure 22(c)** is five times smaller (150 V).

The ability of the SHAPF control system to compensate harmonics after the load change is presented in **Figure 23(a)** (see grid current waveforms I_S). The load DC resistance is changed from 36.5 to 18.25 Ω and the inverter DC voltage is presented in **Figure 23(b)**. Observing the zooms in **Figure 23(a)**, it can be noticed that after the load change, the grid current harmonics are well reduced (THD_{IS} is reduced from 36.65 to 4.6%), but the grid fundamental reactive power (Q_{S1}) is partially compensated (from 2552 to 1336 Var) because of the fixed SHAPF passive part reactive power (1210 Var see **Table 11**). This is one of the disadvantages of such SHAPF topology.

	No filter	n_{re} = 4,87	n_{re} = 6,87	n_{re} = 10,87
$Q_{S1(1)}$ [Var]	1234	−48.7	8,16	18,31
$P_{S1(1)}$) [W]	1435	1438	1437	1435
$Q_{f1(1)}$ [Var]	—	−1283	−1226	−1216
$S_{f1(1)}$ [VA]	—	1282,83	1225,70	1217,54
$P_{load1(1)}$ [W]	—	1437	1437	1437
$Q_{load1(1)}$ [Var]	—	1234	1234	1234

Table 13.
One-phase representation of the fundamental harmonic active, reactive, and apparent powers at the load, SHAPF, and grid side.

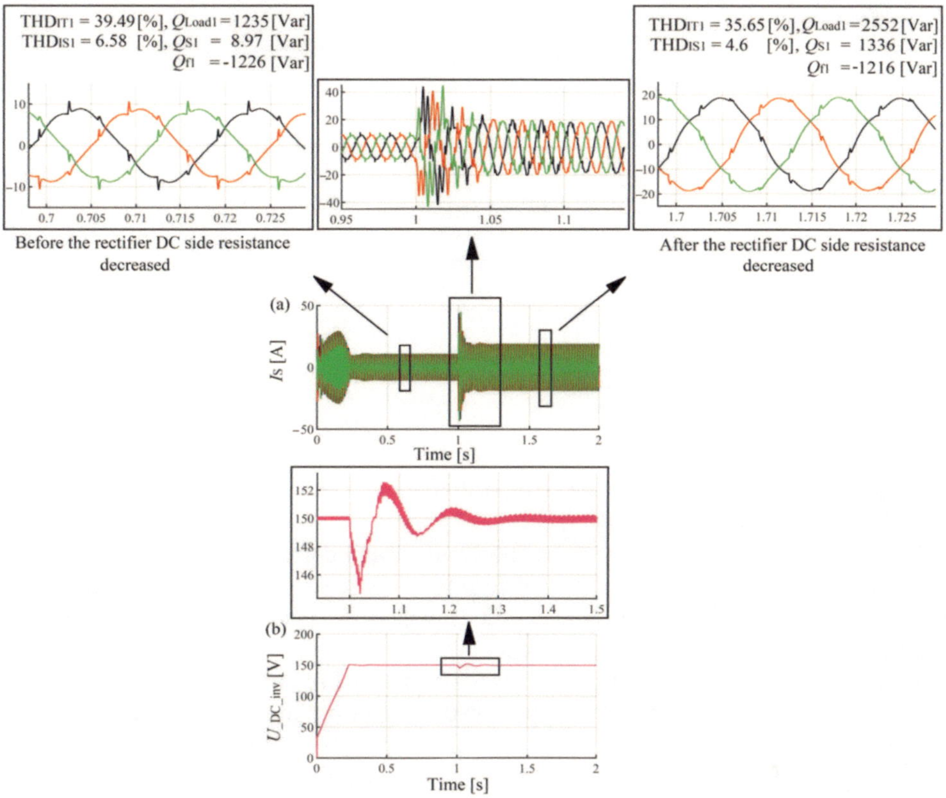

Figure 23.
(a) Grid current waveforms and (b) SHAPF DC voltage before and after the load change (Rdc is changed from 36.5 to 18.25 Ω).

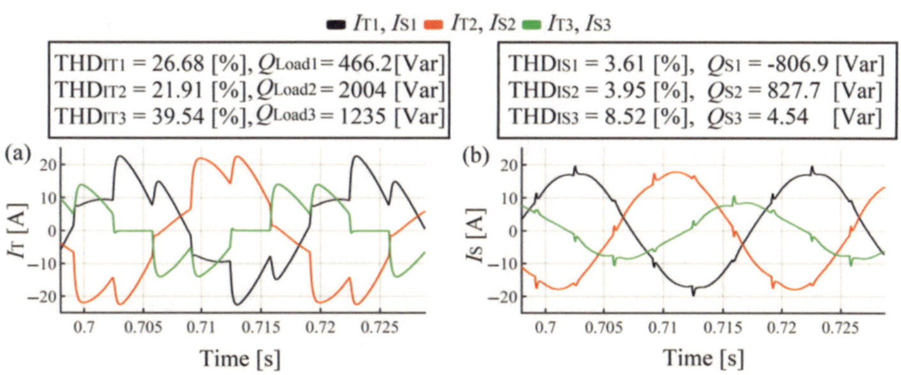

Figure 24.
Unbalance load (k5 closed—see ***Figure 1****): Waveforms of the load current (a) and grid current (b).*

Figure 24(a) and (b) presents the grid and load current waveforms respectively during the asymmetry (k5 closed—see **Figure 1**). It can be noticed that the SHAPF with the proposed control system (see **Figure 18**) does not have ability to compensate the asymmetry component. This is another disadvantage of such SHAPF topology.

The power system impedance versus frequency characteristic observed from the thyristor bridge input is presented in **Figure 25**. On that characteristic, it can be

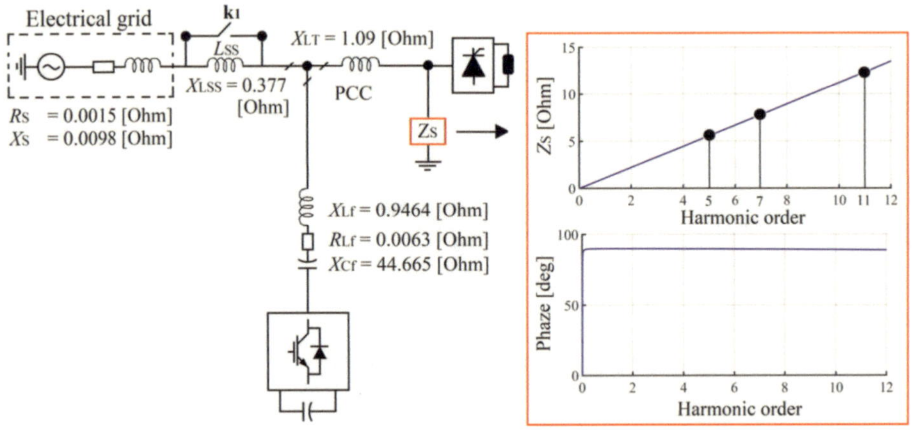

Figure 25.
Frequency impedance characteristic of the power system seen from thyristor bridge input (k1 closed).

noticed no resonance phenomena (parallel and series) contrary to the case of the characteristics in **Figure 7**. The parallel resonance between the PHF (when operating alone) and the electrical gird is eliminated as well as the grid dependency of the PHF work efficiency (when operating alone).

Figure 26 presents the grid current waveforms before and after the SHAPF connection. It can be noticed that the ripples at the high rate of current change (see zoom **Figure 26**) are better mitigated when the passive part is tuned to the frequencies of 243.5 and 543.5 Hz, respectively, near to the frequencies of the 7th and 11th harmonics (see grid current waveforms in **Figure 21**). Observing **Table 11**, it can be seen that with those frequencies, the passive part reactor inductance (L_f) is smaller than the thyristor bridge input reactor inductance (L_T). Therefore, in this case study, the SHAPF passive part tuning frequency should be chosen taking also into account the size of the thyristor bridge input reactor which should be higher for a better ripple mitigation at the commutation points.

The performed studies have shown that when the SHAPF passive part is tuned to the frequency of harmonic higher than the 5th and 7th harmonics: the higher grid current harmonics (e.g., from the 13th) are better reduced because the PHF presents smaller impedance for those harmonics, the 5th and 7th harmonics currents are worse

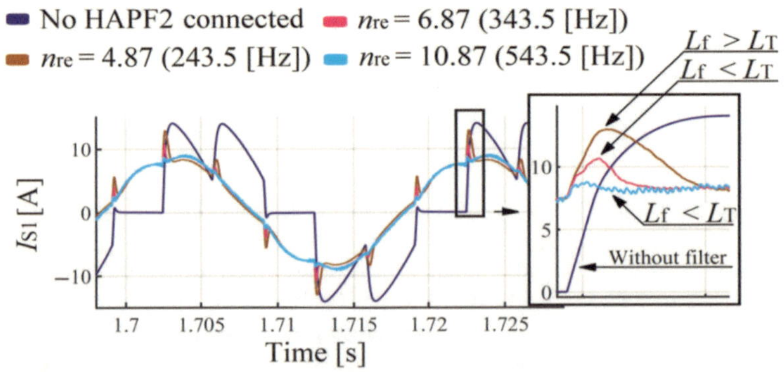

Figure 26.
Waveforms of the grid current before and after the SHAPF connection.

reduced, and the PHF reactor size is smaller therefore low cost. In the case where the passive part reactor size is smaller than the one at the thyristor bridge input, the switching ripple are worst mitigated, but the ripples at the commutation points are better mitigated.

6. Conclusion

The detail investigations on the PHF, SAPF, and HAPF work efficiency have been presented in this chapter. They were performed in the electrical system with six-pulse thyristor-bridge as load. The first investigation concerned the grid dependency of the PHF work efficiency in terms of harmonics mitigation. And it has been recommended the use an additional line reactor between the PCC and the filter to increase the PHF effectiveness in terms of harmonics mitigation.

In the second one, the influence of the thyristor-bridge input reactor size on the SAPF work efficiency was studied. And it has been recommended to choose or compute the SAPF input reactor parameters taking also into account the thyristor bridge input reactor size. According to that investigation, the SAPF input reactor inductance should be equal or smaller than the one at the thyristor bridge input for good work efficiency.

The last investigation was related to the SHAPF, comparing its work efficiency for different tuning frequencies. It has been demonstrated that for a better ripple reduction at commutation points, the passive part tuning frequency should be chosen in such a way that the filter reactor inductance is smaller than the one of the thyristor bridge input reactor. With the inverter connected in series with the PHF, the grid dependency of the PHF work efficiency is eliminated as well as the parallel resonance that could occurred between the grid and PHF. One of the useful advantages in that topology is that the SAPF DC voltage can be considerably reduced in comparison with the case when it is operating alone. The main disadvantages of such topology are that it has fixed compensating reactive power and presents difficulty to compensate asymmetry component.

Conflict of interest

The authors declare no conflict of interest.

Author details

Chamberlin Stéphane Azebaze Mboving* and Zbigniew Hanzelka
Department of Power Electronics and Energy Control Systems, Faculty of Electrical Engineering, Automatics, Computer Science and Biomedical Engineering, AGH University of Krakow, Poland

*Address all correspondence to: stephane@agh.edu.pl

References

[1] EN 50160: voltage characteristics of electricity supplied by public electricity network [Internet]. 2010. Available from: https://standards.iteh.ai/catalog/standards/clc/18a86a7c-e08e-405e-88cb-8a24e5fedde5/en-50160-2010 [Accessed: 2023-05-27]

[2] Nikunj S. Harmonics in power system – Causes, effects and control Siemens Industry, Inc. [Internet]. 2013. Available from: https://assets.new.siemens.com/siemens/assets/api/uuid:8ab2a02e-ad94-41cb-a362-438f016aa704/drive-harmonics-in-power-systems-whitepaper.pdf [Accessed: 2023-05-30]

[3] Wagner VE, Balda JC, Griffith DC, McEachern A, Barnes TM, Hartmann DP, et al. Effects of harmonics on equipment. IEEE Transactions on Power Delivery. 1993;**8**(2):672:680. DOI: 10.1109/61.216874

[4] Kuldeep KS, Saquib S, Nanand VP. Harmonics and its mitigation technique by passive shunt filter. International Journal of Soft Computing and Engineering (IJSCE). 2013;**3**(2):325-332. ISSN: 2231-2307

[5] Bhim S, Ambrish C, Al-H K. Power Quality: Problems and Mitigation Techniques. Southern Gate: Chichester, UK: John Wiley & Sons Ltd, The Atrium; 2015. DOI: 10.1002/9781118922064

[6] Azebaze MCS. Investigation on the work efficiency of the LC passive harmonic filter chosen topologies. Electronics. 2021;**10**:896. DOI: 10.3390/electronics10080896

[7] Azebaze MCS, Hanzelka Z, Firlit A. Analysis of the factor having an influence on the LC passive harmonic filter work efficiency. Energies. 2022;**15**: 1894. DOI: 10.3390/en15051894

[8] Das JC. Passive filters – Potentialities and limitations. IEEE Transactions on Industry Applications. 2004;**40**:232-241. DOI: 10.1109/TIA.2003.821666

[9] Dekka AR, Beig AR, Poshtan M. Comparison of passive and active power filters in oil drilling rigs. In: IEEE International Conference on Electrical Power Quality and Utilization (EPQU). Lisbon, Portugal: IEEE; 17-19 October 2011. DOI: 10.1109/EPQU.2011.6128815

[10] Young-Sik C, Hanju C. Single-tuned passive harmonic filter design considering variances of tuning and quality factor. Journal of International Council on Electrical Engineering. 2011; **1**:7-13. DOI: 10.5370/JICEE.2011.1.1.007

[11] Abdelmadjid C, Jean-paul G, Fateh k. On the design of shunt active filter for improving power quality. In: IEEE International Symposium on Industrial Electronics. Cambridge, UK: IEEE; 18 November 2008. pp. 1020-1025. DOI: 10.1109/ISIE.2008.4677277

[12] Chiang SJ, Chang JM. Design and implementation of the parallelable active power filter. In: 30th Annual IEEE Power Electronics Specialists Conference. Charleston, SC, USA: IEEE; 1999. pp. 406-411. DOI: 10.1109/PESC.1999.789037

[13] Sut-Ian H, Chi-Seng L, Man-Chung W. Comparison among PPF, APF, HAPF and a combined system of a shunt HAPF and a shunt Thyristor controlled LC. In: IEEE Region 10 Conference TENCON. Macao, China: IEEE; 1-4 November 2015. DOI: 10.1109/TENCON.2015.7373058

[14] Kwak S, Toliyat HA. Design and rating comparison of PWM voltage source rectifiers and active power filters for AC drives with unity power factor. IEEE Transaction on Power Electronics. 2005;**20**. DOI: 10.1109/TPEL.2005.854055

[15] Dixon JW, Contardo JM, Moran LA. A fuzzy-controlled active front-end rectifier with current harmonic filtering characteristics and minimum sensing variables. IEEE Transactions on Power Electronics. 1999;**14**. DOI: 10.1109/63.774211

[16] Singh B, Verma V, Chandra A, Al-Haddad k. Hybrid filters for power quality improvement. In: IEE Proceedings – Generation, Transmission and Distribution. Vol. 152. 2005. DOI: 10.1049/ip-gtd:20045027

[17] Francesco G, Andrea F, Alessandro U. New harmonics current mitigation technique in induction motor driving reciprocating compressor. In: IEEE International Symposium on Systems Engineering. Rome, Italy: IEEE; 28-30 September 2015. DOI: 10.1109/SysEng.2015.7302737

[18] Hussein AK. Harmonic mitigation techniques applied to power distribution networks. In: Advances in Power Electronics. Vol. 2013. Hindawi Publishing Corporation; 2013. Article ID: 591680. DOI: 10.1155/2013/591680

[19] Dawid B, Jarosław M, Zygmanowski M, Adrikowski T, Grzegorz J, Jeleń M. Control strategy of 1 kV hybrid active power filter for mining applications. Energies. 2021;**14**:4994. DOI: 10.3390/en14164994

[20] Zhaoxu L, Mei S, Jian Y, Yao S, Xiaochao H, Josep MG. A repetitive control scheme aimed at compensating the 6k+1 harmonics for a three-phase hybrid active filter. Energies. 2016;**9**:787. DOI: 10.3390/en9100787

[21] Azebaze MCS, Firlit A. Investigation of the line-reactor influence on the active power filter and hybrid active power filter efficiency: Practical approach. Przegląd Elektrotech. 2021;**97**: 39-44. DOI: 10.15199/48.2021.03.07

[22] Firlit A, Kołek K, Piątek K. Heterogeneous active power filter controller. In: Proceedings of the IEEE International Symposium ELMAR. Zadar, Croatia: IEEE; 18–20 September 2017. DOI: 10.23919/ELMAR.2017.8124477

[23] Akagi H, Tamai Y. Comparisons in circuit configuration and filtering performance between hybrid and pure shunt active filters. In: IEEE Conference Record of the Industry Applications, 38th IAS Annual Meeting. Salt Lake City, UT, USA: IEEE; 12-16 October 2003. DOI: 10.1109/IAS.2003.1257702

[24] Fujita H, Akagi H. A practical approach to harmonic compensation in power systems - series connection of passive and active filters. IEEE Transaction on Industry Applications. 1991;**27**. DOI: 10.1109/28.108451

[25] Gutierrez B, Kwak S-S. Finite set model predictive control method of shunt hybrid power filter. In: IEEE 9th International Conference on Power Electronics and ECCE Asia, Seoul, Korea. Seoul, Korea (South): IEEE; 1-5 June, 2015. DOI: 10.1109/ICPE.2015.7168176

[26] Kedra B. Comparison of an active and hybrid power filters devices. In: IEEE 16th International Conference on Harmonics and Quality of Power (ICHQP). Romania, Bucharest: IEEE; 25-28 May 2014. DOI: 10.1109/ICHQP.2014.6842771

[27] Azebaze MCS, Hanzelka Z. Hybrid power active filter - effectiveness of passive filter on the reduction of voltage and current distortion. In: Proceedings of the IEEE International Conference on Electric Power Quality and Supply Reliability, Estonia, Tallinn. Tallinn, Estonia: IEEE; 29-31 August 2016. DOI: 10.1109/PQ.2016.7724095

[28] Akagi H. Modern active filters and traditional passive filters. Bulletin of the Polish Academy of Sciences, Technical Sciences. 2006;**54**:3

[29] Mustapha S, Gaubert J-P, Chaoui A, Krim F. Control strategy of a transformerless three phase shunt hybrid power filter using a robust PLL. In: IECON 2012 - 38th Annual Conference on IEEE Industrial Electronics Society. Montreal, QC, Canada: IEEE; 25-28 October 2012. pp. 1258-1267. DOI: 10.1109/IECON.2012.6388557

[30] Srianthumrong S, Akagi H. A medium voltage transformerless AC/DC power conversion system consisting of a diode rectifier and a shunt hybrid filter. IEEE Transactions on Industry Applications. 2003;**39**:3. DOI: 10.1109/TIA.2003.811787

[31] Abu-Rubu H, Iqbal A, Guzinski J. High Performance Control of AC Drives with MATLAB/SIMULINK Models. Chichester, United Kingdom: John Wiley & Sons; 2012. DOI: 10.1002/9781119969242

[32] Moran LM, Dixon JW, Wallace RR. A three-phase active power filter operating with fixed switching frequency for reactive power and current harmonic compensation. IEEE Transactions on Industrial Electronics. 1995;**42**:402-408. DOI: 10.1109/41.402480

[33] Akagi H, Kanazawa Y, Fujita K, Nabae A. Generalized theory of instantaneous reactive power and its application. Electrical Engineer in Japan. 1983;**103B**. DOI: 10.1002/eej.4391030409

Chapter 6

The Hoisting Machines as Source of Higher Harmonics in Underground Mines

Tomasz Siostrzonek

Abstract

The chapter describes the structure of drive systems used in hoisting machinery of underground mines. These devices, due to their special application and power, are characterized by special traits. The most commonly used systems with DC reciprocating motors are described. The so-called complex (multi-pulse) converters are used in those solutions. An extra attention was paid to the impact of these systems on the power supply network, especially the generation of higher harmonics. The results of measurements made under real conditions are presented. Measurements were made before and after the installation of new solution in connection with the modernization of systems due to the need to increase the efficiency of these machines.

Keywords: power converters, hoisting machines, multi-pulse system, power quality, DC drive systems, power grids in undergroud mines

1. Introduction

To analyze higher harmonics as one of the elements of the quality of electricity supply, it is necessary to clarify what their sources are and how it is possible to describe waveforms containing higher harmonics.

Higher harmonics—such a term is used to describe waveforms with frequencies that are integer multiples of the fundamental frequency. If we assume that the frequency of the fundamental waveform is 50 Hz, then the harmonics will have frequencies as given in **Table 1**.

Sources of higher harmonics are loads with nonlinear current-voltage characteristics. Such a load, connected to the power grid, causes a current flow in the network with a distorted waveform. The flow of non-sinusoidal current causes voltage drops on the network reactances, which are not sinusoidal waveforms, although the source voltage has a sinusoidal waveform.

Analysis of distorted waveforms is possible by using Fourier series [1–7]. A periodic, distorted time waveform of voltage or current can be decomposed into sinusoidal waveforms of different frequencies. Eq. (1) is the Fourier series for the function f(t) [8]:

Harmonic order (h)	Harmonic frequency [Hz}
2	100
3	150
4	200
5	250
...	...
n	n*50

Table 1.
Summary of harmonic orders and their frequencies.

$$f(t) = \frac{a_0}{2} + \sum_{h=1}^{\infty} a_{(h)} \cos(h\omega t) + \sum_{h=1}^{\infty} b_{(h)} \sin(h\omega t) \tag{1}$$

Where:

$$a_0 = \frac{1}{\pi} \int_0^{2\pi} f(t) dt \tag{2}$$

$$a_{(h)} = \frac{1}{\pi} \int_0^{2\pi} f(t) \cos(h\omega t) dt \tag{3}$$

$$b_{(h)} = \frac{1}{\pi} \int_0^{2\pi} f(t) \sin(h\omega t) dt \tag{4}$$

In another form, the Fourier series can be written:

$$f(t) = \frac{a_0}{2} + \sum_{h=1}^{\infty} A_{(h)} \sin(h\omega t + \varphi_h) \tag{5}$$

The components of the sum in Eq. (5) are called harmonics of the periodic waveform f(t).

Distorted waveforms are a very big problem in power grids. Assessing the degree of distortion and defining acceptable distortion levels is a major challenge. The most well-known indicator for determining the level of distortion is total harmonic distortion (THD). It will be described in detail in the following section.

2. The power grids of mines

The subject described here concerns phenomena related to higher harmonics in the electrical networks of underground mines in Poland. According to data published by the State Mining Authority, there are currently 36 underground mines in Poland [9]. These include coal mines, ore mines, and salt mines (**Table 2**).

Each of the aforementioned types of mines has different characteristics of work, and consequently, the electrical networks of these plants are subject to different disturbances. However, it is possible to find some common features that are present in

Underground mining facilities extracting	Number of mines
Coal	20
Coal (mines in decommissioning)	8
Coal (mines under construction)	1
Copper ores	3
Zinc and lead ores (mines in decommissioning)	1
Salt	1
Gypsum and anhydrite	1
Brine for therapeutic purposes	1

Table 2.
Number of mines in Poland [9].

all mining plants, as they are subject to the same regulations. These include regulations on crucial equipment, the operation of which affects the safety of crews and equipment.

The power supply of underground mines is carried out in accordance with the principle of its reliability; that is, each mine must have a power supply from two independent sources of electricity supply. In most mines, the power supply is provided by double overhead or cable lines with a rated voltage of 110 or 220 kV. Due to the fact that the decommissioning of mines in Poland has been progressing in recent years, some mines have a medium-voltage power supply in addition to the meddium voltage (MV) level. However, MV power supply is a backup power supply, due to the impossibility of powering all technological equipment, which power exceeds a dozen megawatts. The use of a power supply from the medium-voltage level as a backup power supply is in accordance with current regulations, which specify that the second power supply must have ensure parameters that it is possible to run devices that guarantee the safety of people, equipment, and the environment. This should be understood in such a way that the backup power supply is sufficient, in terms of parameters, for the operation of equipment used to evacuate the crew and ensure the operation of the main dewatering pumps and the main ventilation fans of the mine.

Like any industrial plant, in a mine, special attention is paid to the correct operation of equipment because only in this case it is possible to ensure adequate economic performance and safety of workers. According to Ref. [10], in every mining, there are so-called basic facilities, that is, those which correct operation guarantees the safety of the crew and equipment. Equipment that are elements of these facilities should have two independent sources of power supply. Based on Article 29 [10], basic facilities include:

1. shafts with equipment;

2. hoisting machines in shafts;

    a. stations:

    b. main fans,

    c. de-methanation together with a network of pipelines;

3. headquarters of the company-wide telephone communications system, dispatching centers of the traffic dispatcher's systems, mining geophysics stations, and trunk telecommunications networks;

4. main air compressor stations, together with a network of shaft pipelines;

5. main drainage facilities and systems;

6. main depots for fuels, oils, and lubricants, as well as permanent fuel-filling chambers for transport vehicles;

7. main facilities for producing and transporting backfill and sealing mixtures;

8. stationary air-conditioning equipment with a nominal cooling capacity of more than 1 MW;

9. transport equipment, the means of transport of which move on a track with an inclination of more than 45°, in mine workings;

10. high-voltage and medium-voltage electrical equipment, installations, and networks supplying the facilities, machinery, and equipment referred to in items.

The permissible duration of interruption of power supply to these devices is determined by the Mine Site Operation Manager, and each time results from the specific conditions in the mine. For example, if the main ventilation fan is damaged and the standby fan cannot be started, underground work is halted, electrical equipment in the relevant methane fields (if applicable) is switched off, and the evacuation of the crew begins. The special, detailed regulations for basic equipment testify to a very complex problem when it comes to powering the mine and individual equipment. It is also related to the fact that special attention is paid to the problems of interference in the networks of mining plants and the consequences it can have on the safety of the working crew. In each underground mine, the surface and underground parts of the power grid can be distinguished. In each case, the network layouts of the surface and underground parts are different.

Figure 1 shows a block, simplified diagram of the power grid of a mining plant. The dashed line indicates the backup power supply. There may be situations where a mining plant has three power supply points.

From the medium-voltage switchgear in the surface section, the most important primary facilities are supplied: hoisting machines and fan stations. Two types of network layouts are used in mine networks: IT and TN. The insulated network is used to supply power to equipment operating underground. TN systems are used in networks on the surface. This is important for further consideration because when supplying three-phase equipment from a network with an isolated zero point, there is no zero component, that is, harmonics, the order of which is an integer multiple of 3 in the expansion of the function into Fourier series. One of the criteria for evaluating multi-pulse power systems for hoisting machinery is the level of their negative impact on the power supply network. In particular, it refers to the distortion of voltage due to the flow of distorted current. The legislation [11] defines numerical measures for dimensioning this impact. The basic coefficients used to assess voltage distortion are the total harmonic distortion (THD) and the values of individual harmonics [12–15].

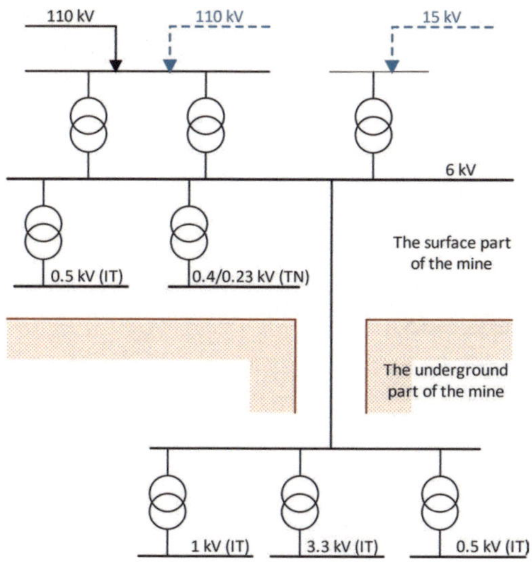

Figure 1.
The example diagram of a power grid in a mine.

The evaluation is carried out on the medium-voltage side (on the primary side of the converter transformer). Regulations [11] strictly define permissible levels of harmonics and the value of the voltage THD.

3. The sources of higher harmonics in networks of mining plants

In the power grids of mines, the sources of higher harmonics are primarily power electronic converters used to supply the drives of various equipment. Such equipment includes:

1. Hoisting machines,

2. Fans,

3. Drives for conveyors and pumps.

AC/DC converters, AC/DC multi-pulse systems, AC/DC/AC intermediate frequency converters, and soft-starts are used in the mentioned equipment. The hoisting machine in any mine is the largest load and the largest source of interference. The power of hoisting machines ranges from several hundred kilowatts to several megawatts. In Poland, the most common machines range from 1.5 to 7 MW. In comparison, the powers of the main fan motors range from 900 kW to 1.5 MW. In addition, the fan operates in a continuous mode in contrast to the exhaust machine, as will be discussed later in the chapter. Other receivers pose a potential threat, but the magnitude of this threat is determined by the number of these devices, which varies from mine to mine, and the simultaneity of operation.

3.1 The hoisting machine

A hoisting machine transports ore, materials, and crew from underground mines to the surface. It is usually the only transport route from underground workings. That is why it is so important for these machines to work properly.

The hoisting machine is an electromechanical system that includes:

- Koepe pulley or drum,
- hoisting vessels (cage, skip),
- ropes,
- motor,
- converter together with a transformer,
- machine control system,
- braking system,
- signaling and shaft communication.

Figure 2 shows a diagram of a winding machine with two types of linkage: a Koepe wheel (**Figure 2a**) and a drum (**Figure 2b**).

In the machine with the Koepe pulley, the support rope is wound through a special type of wheel. The wheel is equipped with special grooves with a lining to increase the coefficient of friction. For proper operation of the system, it is necessary to use an equalizing (tail) rope to bring the system into full balance. Failure to use an equalizing rope would result in the need for a much more powerful motor, and the entire system would be subject to the occurrence of slippage of the rope relative to the drive wheel. The use of an equalizing rope means that the size of the drive motor is selected only because of the weight of the load being carried in the hoisting vessel. This mass determines the static overweight that occurs in this kinematic system.

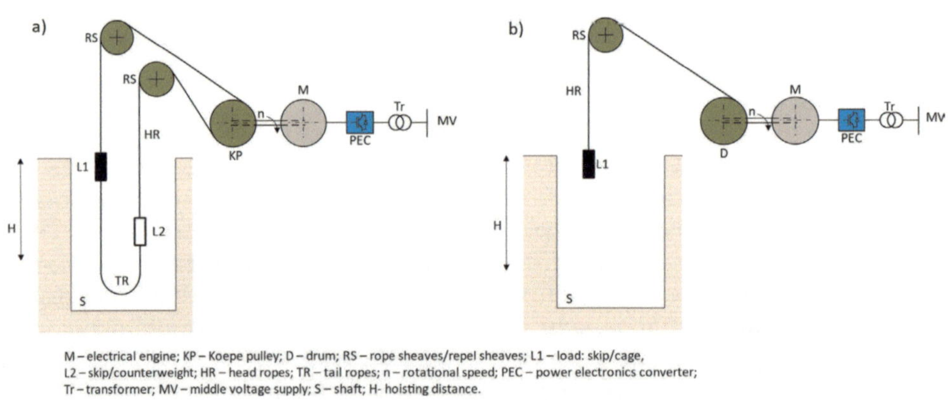

Figure 2.
Block diagram of the hoisting machine. (a) with Koepe pulley, (b) with drum.

The second type of machine, shown in **Figure 2b**, is a drum machine. In this case, the rope is wound on a drum.

Regardless of the type of executive system implementing the displacement of vessels in the shaft, the following are used to drive the rope drive:

- separately excited DC motors (**Figure 3a**),
- squirrel-cage induction motors,
- slow-rotating synchronous motors (**Figure 3b**).

You can still find working systems of hoisting machines with slip-ring motors. The speed of these motors is controlled through attached resistances in the rotor circuit. For economic reasons, such solutions are no longer used.

3.2 The operation of the winding machine: Driving diagram

The operation of a hoisting machine is cyclic. The duty cycle of a hoisting machine consists of three stages: acceleration, travel with fixed speed, and deceleration. The length of the driving cycle, and thus the lengths of its individual parts, depends on the depth of the shaft, driving speed, acceleration, and deceleration, and on the mode in which the machine is used, that is, whether it is to transport ore, materials, or people. **Figure 4** shows an example of a seven-period driving diagram. This is a driving diagram for a system with two vessels (skips) for transporting ore from mining level. The diagram is seven-period because it still takes into account additional stages, where the vessel moves in special guidance systems so that precise travel to the skip filling or emptying point is possible. It differs slightly from the diagram for a machine used to transport people. For the transportation of people, the vessel cannot move faster than 12 m/s. (respectively 43.2 km/h). This restriction does not apply to machines transporting materials or excavated material.

The time of the entire cycle ranges from tens of seconds. For a machine in which the vessels move at a speed of 12 m/s, the acceleration and deceleration is 1.0 m/s^2, and the travel distance is equal to 500 m the duration of one cycle is about 80 seconds (including the time required for loading and unloading).

(a) (b)

Figure 3.
The winding machine with a DC motor (a) and a synchronous motor (b). [SIEMAG TECBERG POLSKA].

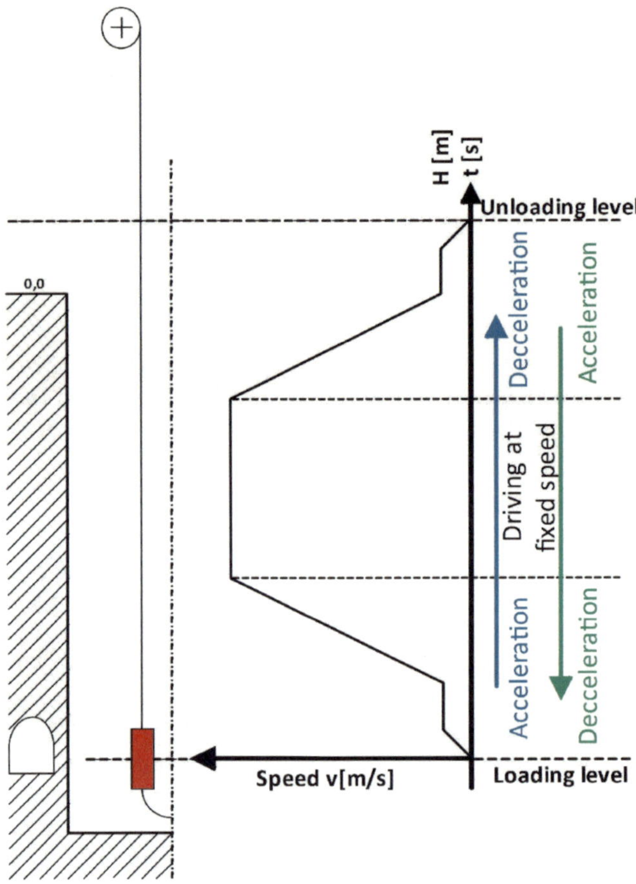

Figure 4.
The hoisting diagram.

Therefore, this type of high-power device is very troublesome with regard to the formation of disturbances in the power grid. The load is cyclically variable, depending on the load—that is, on the mass of excavated material poured into the vessel or the number of people in the cage. The timing of the start in the excavated material conveying equipment depends on the quality of the excavated material and affects the start and end of the entire cycle.

It is not possible to synchronize the operation of the machine with other equipment in the mine, since it is not possible to precisely determine the moment of start and end of the cycle for operating hoisting machines.

3.3 The hoisting machine with the separately excited DC motor

Most Polish mines (more than 90%) use a slow-speed DC motor (gearless solution). Due to its simplicity and linear dependencies, the mathematical description has been available in the literature for years [16]. More relevant is such a power supply for a reciprocating motor that allows smooth speed control from 0 to its rated value. Several decades ago, this was reached through the so-called Leonard circuit. This is an electromechanical system that provides smooth speed control over the entire control range. In the 1970s, static thyristor inverters were introduced. Power supply to the

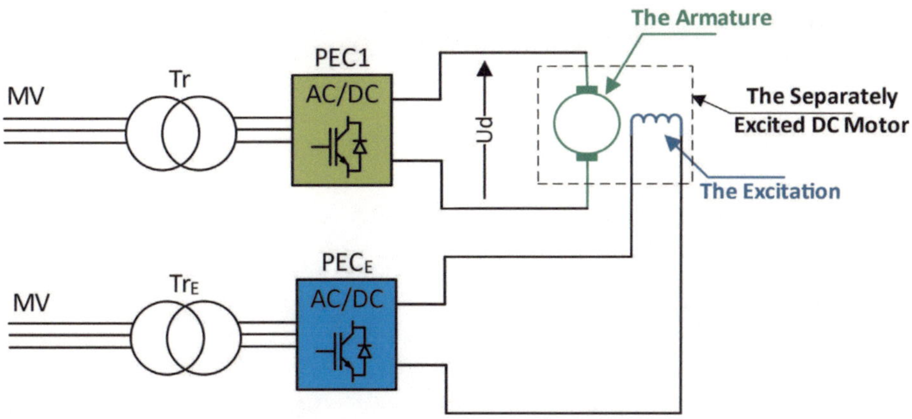

Figure 5.
Basic power supply diagram of separately excited DC motor; MV—Middle voltage source, tr—Main transformer (form armature), TrE—Excitation transformer, PEC1—Power electronic converter, PEC_E—Power electronic converter for excitation.

motor was provided by two controlled rectifier systems (**Figure 5**): PEC1—the system supplying the motor armature (also known as the main circuit of the winding machine), and PEC_E—the rectifier supplying the motor excitation.

The direction of rotation of the machine can be changed by inversing the direction of current flow in the excitation circuit or in the armature circuit. Due to the cost of four-quadrant converters, which depended on their power, the change of direction of rotation by changing the direction of current flow in the excitation was used. This operation is subject to very strict control, due to the fact that a decrease in motor excitation current causes an increase in speed. In a hoisting machine, speed control is achieved by changing the armature voltage. The system is equipped with two interrelated control loops: voltage regulation and excitation current regulation. When the direction of rotation is changed, if the system has not managed to reach zero speed, with a reduction of the excitation current. The armature supply voltage is also reduced.

This problem does not occur in machines equipped with a four-quadrant converter in the armature circuit. The direction of rotation is altered by changing the direction of current flow in the armature. The excitation circuit is supplied from a single-quadrant rectifier, so there is no danger of de-energizing the machine.

A serious problem, from the point of view of the impact on the power grid, is the rectifier of the armature circuit. **Figure 6** shows the supply voltage waveforms and the phase current waveform for a 6-pulse bridge. Based on Ref. [8], the THD_I harmonic content factor for a three-phase thyristor bridge converter is determined by Eq. (6). The equation does not take into account the effect of commutation and the variable component of the load current:

$$THD_I = \sqrt{\left(\frac{I}{I_1}\right)^2 - 1} \approx 0,31 \qquad (6)$$

where:
I—root mean square (RMS) value of the phase current,
I_1—RMS value of the fundamental component of the phase current.

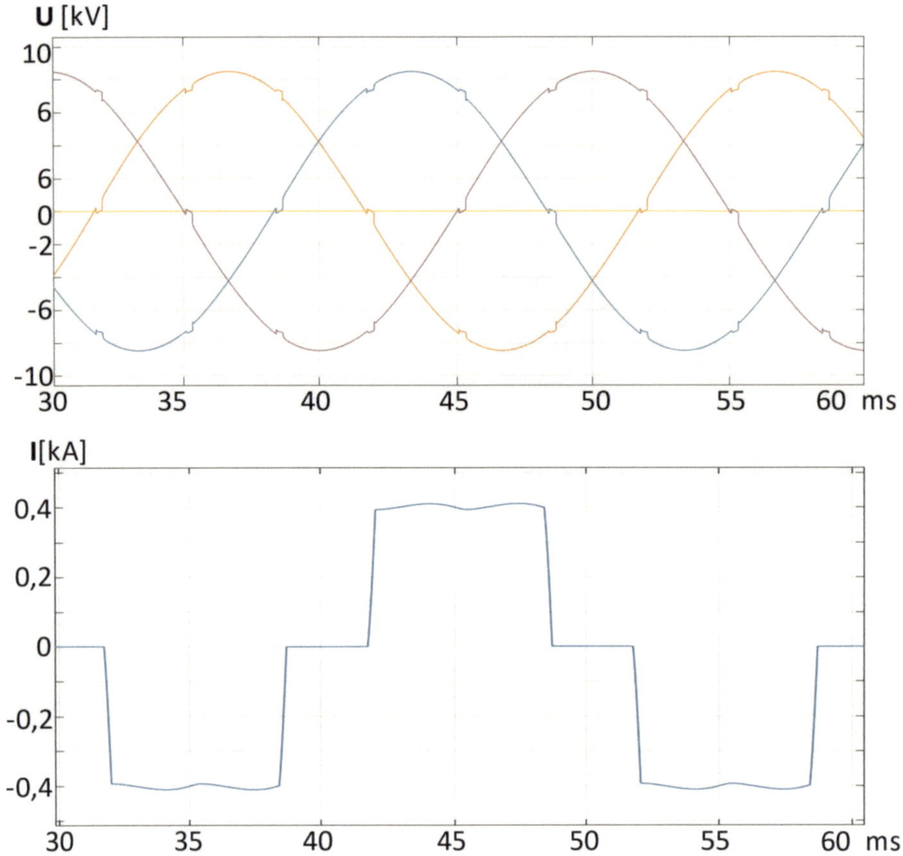

Figure 6.
The waveform of the supply voltage and phase current of a 6-pulse bridge rectifier.

Figure 7 shows the current harmonic spectrum of a 6-pulse bridge system—the results of simulation studies. According to Eq. (7), there are characteristic harmonics in the current waveforms depending on the number of pulses of the converter:

$$h = 6m \pm 1 \tag{7}$$

where:

m = 1,2, ... n.

h – harmonic order.

In such system, the harmonics fifth and seventh are the most significant for the distortion of the power grid waveforms. Their value in relation to the fundamental harmonic is 20 and 14%, respectively.

The current harmonic content factor is almost 0.29.

The impact of a high-power 6-pulse system is a problem in terms of the proper operation of the power grid. Therefore, complex converters are used to power high-power drives. DC converters containing two thyristor bridges connected in series are used in hoisting machines. This eliminates connecting thyristors in series to achieve higher output voltages. **Figure 8** shows a schematic of the power supply to the armature circuit of a DC hoist motor through a compound converter.

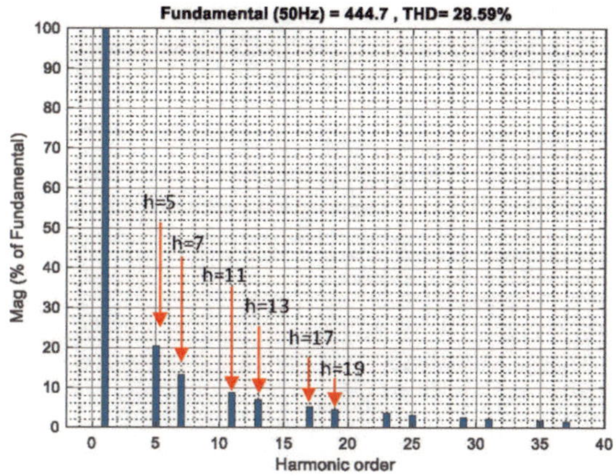

Figure 7.
The harmonic current spectrum of a three-phase 6-pulse bridge system.

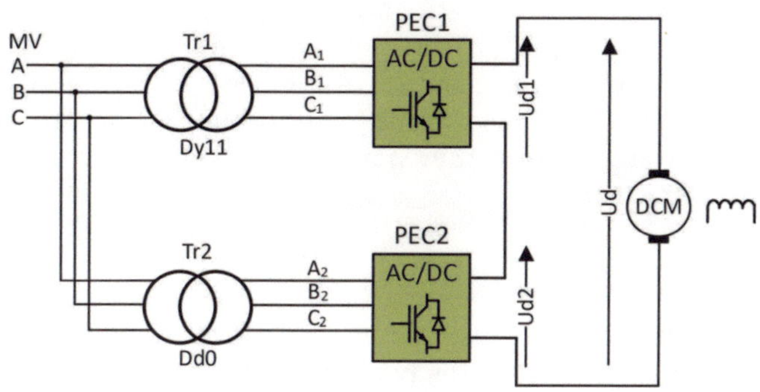

Figure 8.
Diagram of the 12-pulse circuit.

Two bridge converters are connected in series on the output voltage side. The voltage supplying the armature circuit of the hoisting motor is the sum of the output voltages of the individual bridges. Each of the bridges is powered by a converter transformer. The transformer can be of a three-winding design. Two transformers are used in hoisting machine drives for safety reasons. Such a solution allows the drive to operate in emergency conditions. Damage to one of the converters or transformers allows the motor to operate with limited parameters but does not completely eliminate the drive. It ensures safe evacuation of people during the failure of certain elements of the hoisting machine. The converter transformers are powered from a medium-voltage line, and their connection groups shift the phase-to-phase voltages of the secondary sides of the transformers by 30° with respect to each other (**Figure 9**).

The RMS value of the harmonic h of the source phase current is [8]:

$$I_{hz} = 2I_h\sqrt{\frac{1+(-1)^m\cos(h\Delta\alpha)}{2}} \tag{8}$$

where:

$$\Delta\alpha = (\alpha_1 - \alpha_2), \tag{9}$$

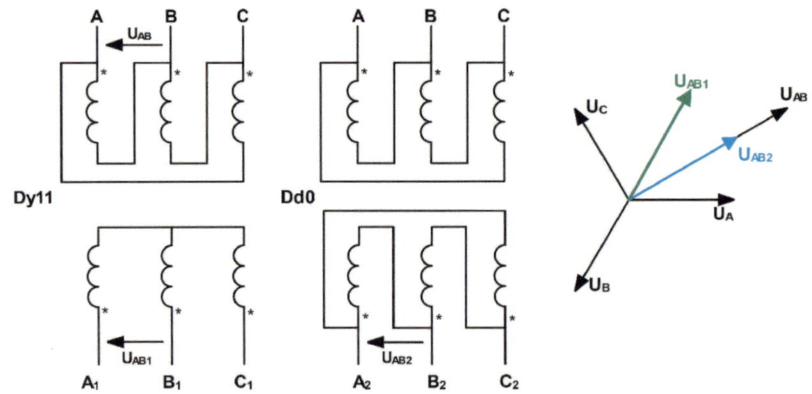

Figure 9.
The configuration of converter transformers.

$$I_h = \frac{\sqrt{6} I_d}{h\pi} \tag{10}$$

For m = 2 k-1, where k = 1,2,3,.... and for $\alpha_1 = \alpha_2$, the I_{hz} of relation (8) is 0. Thus, a converter supplied by transformers with different connection groups and controlled in such a way that the angles of the thyristors' switching delays are equal will produce harmonics of the source current:

$$\mathrm{h} = 12\mathrm{m} \pm 1 \tag{11}$$

That is, this converter is a 12-pulse system, and the control implemented to ensure the equality of the driving angles of the two converters is common.

A characteristic of these converters is that there is an alternating component in the voltage on the DC side with a frequency 12 times higher than the frequency of the supply voltage. It is therefore lower than when using a single hexapulse rectifier. Theoretically, in this system, the lowest of the higher harmonics is the 11th harmonic. The use of common control means that there are no harmonics with strictly defined frequencies in the phase currents of the supply line. Only the so-called characteristic harmonics occur.

There is still a second type of control of complex converters. This is sequential control. It consists of the operation of one of the bridges with a full overdrive, that is, with the maximum conduction angle of the thyristors. At this time, the other converter operates with a variable switching angle of the thyristors. This type of control causes the entire converter to operate as a 12-pulse converter only for short periods of time. In addition, its operation is perceived by the grid as a 6-pulse system, resulting in an increased impact of the system on the power grid. Therefore, common control is used in most cases.

Figure 10 shows the harmonic spectrum of the phase current of a 12-pulse converter. The THD of the current of this converter has decreased compared to the 6-plus converter to about 6%.

4. The systems with an increased number of pulses and less impact on the power grid

In order to further eliminate higher harmonics generated by the converter system, the number of pulses of the system can be increased. It is possible to build DC motor

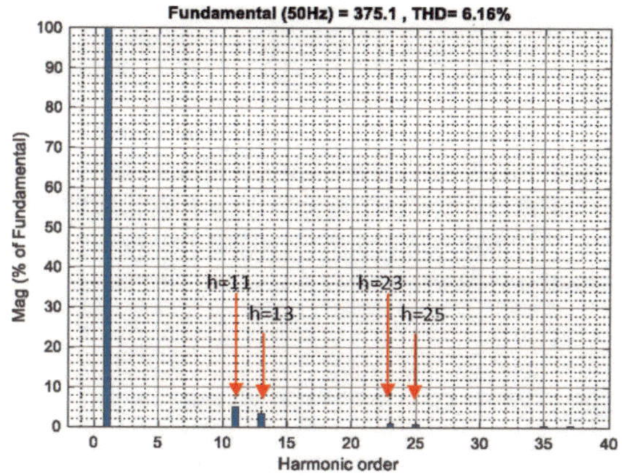

Figure 10.
Harmonic current spectrum of a three-phase 12-pulse system.

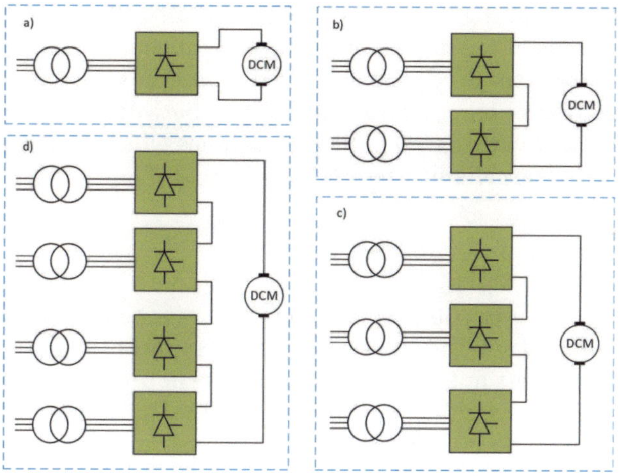

Figure 11.
The configurations of multi-pulse systems: (a) 6-pulse, (b) 12-pulse, (c) 18-pulse, (d) 24-pulse.

power systems with a number of pulses greater than 12. By adding more 6-pulse converters and appropriately configuring power transformers, systems with a number of pulses of 18, 24, and more are built. Such solutions are used, for example, in traction systems. However, in the mine, where unification of solutions is required, and, above all, their simplicity and economic efficiency, such systems are exceptions. **Figure 5** shows diagrams of multi-pulse systems formed by a suitable combination of basic—6-pulse systems (**Figure 11**).

For systems with different numbers of pulses, the characteristic harmonics are defined as follows [5]:

- for a 6-pulse system: h = 6 m ± 1,
- for a 12-pulse system: h = 12 m ± 1 (for common control),

- for an 18-pulse system: h = 18 m ± 1,
- or a 24-pulse system: h = 24 m ± 1,

where m = 1, 2, 3, ...

In addition to the characteristic harmonics for a given type of compound converter, there are also harmonics of orders other than those specified, but their value relative to the basic harmonic is small.

5. The higher harmonics in real 12-pulse systems

The investment cost of building a hoisting machine is significant in the overall cost of operating a mine. However, it is not as high as the cost of a longwall complex used for mining. The product life for a hoisting machine is estimated to be 20–30 years. However, it happens that they work much longer. Hoisting machines operating in Polish mines undergo modernization processes after about 40 years of operation. Due to the current situation of the coal sector, the cost of modernization is reduced each time, and the problems of the impact of the hoisting machine drive system on the mine's power grid are secondary problems. This often leads to major problems related to the proper operation of other equipment in these networks. As an example, the results of measurement work carried out at two mines where hoisting machine upgrades were carried out will be presented.

5.1 The modernization of the hoisting machines

The purpose of modernization, as in many cases, was to increase the mining capability of a mining plant. The hoisting machine is one of the most important components of the process line. The correct production process depends on the performance of the hoisting machine. In this case, in order to ensure that the hoisting machine was not the weakest component, a major modernization, or rather a replacement with a new one, was necessary.

5.1.1 The modernization number 1: The conditions before modernization

The parameters of the system in operation before the upgrade are shown in **Table 3**.

The machine was built as a two-bulb, two-stroke machine with a 5 m diameter Koepe wheel. The 1450 kW slow-speed, reciprocating motor was powered by thyristor converters. **Figure 12** shows an overview diagram of the system.

The power supply to the hoisting motor was implemented through two converter transformers of 1 MVA, each with a voltage of 6/0.4 kV and connection groups Yyn0 and Dyn5. This was a series connection of two six-pulse systems. This resulted in a twelve-pulse effect on the power grid. The change in the direction of rotation of the motor occurred as a result of changing the direction of current flow in the excitation circuit. To make this possible, the excitation circuit converter was bidirectional, i.e., constructed from two reciprocating bridges controlled accordingly. Thanks to this design, the entire drive was able to operate at full assumed load and full draw speed. If one of the main circuit converters failed, it was possible to operate at half speed and

Lp	Parameter	Wartość
1	Motor type	PW-101
2	Power	1450 kW
3	Supply voltage	800 V
4	Rated current	2000 A
5	Excitation current	144 A
6	Rotational speed	61 obr/min
7	Linear speed	16 m/s
8	Mass in the skip	7,5 Mg
9	Depth (drive way)	489 m

Table 3.
Parameters of the machine before modernization.

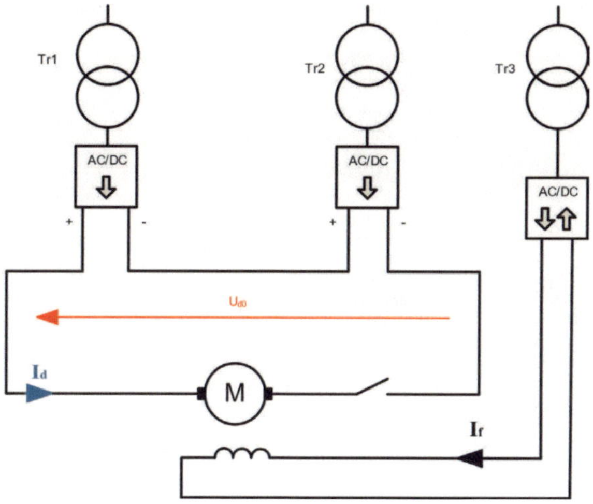

Figure 12.
Diagram of the drive before the upgrade.

full load. The system then operated as a six-pulse system. The switching of the main circuit was provided by a special switching circuit, the so-called switcher.

5.1.2 The modernization number 1: The conditions after modernization

After the upgrade, a DC hoisting motor powered by a system of static converters in the armature and excitation circuits was used. Due to the need for special operations when changing ropes and due to the increase in the weight that can be transported by skip, the rated power of the hoisting motor was increased by 850 kW. The inverse of the rotational speed is changed by the direction of current in the armature circuit. This is a major difference from the system before the upgrade. Now, a non-reversing converter is used in the excitation circuit. A diagram of the solution is shown in **Figure 13**, and the parameters of new machine are summarized in the **Table 4**.

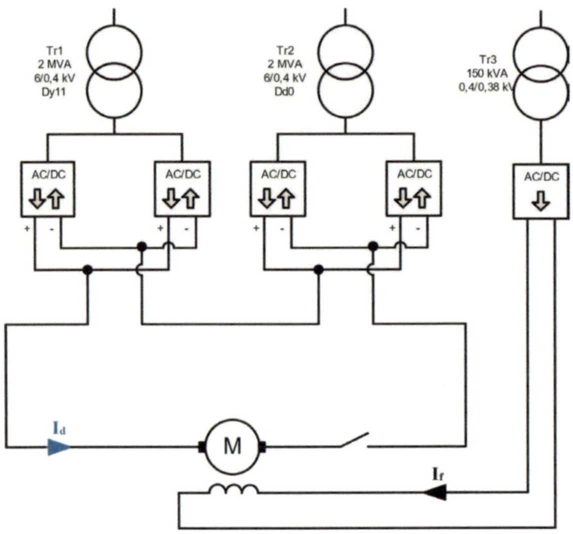

Figure 13.
The diagram of the power circuit of the hoisting machine after modernization.

Lp	Parameter	Wartość
1	Motor type	PW-124
2	Power	2200 kW
3	Supply voltage	853 V
4	Rated current	2810 A
5	Excitation current	225 A
6	Rotational speed	54,57 obr/min
7	Linear speed	12 m/s
8	Mass in the skip	12,5 Mg
9	Depth (drive way)	489 m

Table 4.
The parameters of the machine after modernization.

The armature circuit is powered through four converters in a series-parallel arrangement. This ensures that the required current is achieved (parallel connection of two converters in each branch). Each pair of parallel-connected converters is supplied from a separate converter transformer with the appropriate connection group. The series connection of pairs of converters provides an adequate level of voltage supply to the motor, as well as implements the 12-pulse effect of the drive on the power grid – reduces the magnitude of low-order harmonics, especially fifth and seventh. In the event of a converter failure, the system can operate as 6-pulse. The speed of the drive is then reduced to half the rated speed, which is the result of halving the supply voltage.

Prior to the upgrade, the hoisting engine was already powered by static converters that interacted with the mine plant's network. In such cases, in order to determine whether the new system would work properly without generating additional interference, it was necessary to take measurements of the operating machine and the new hoisting machine after its installation.

Figures 14 and **15** show the locations of the measuring instruments. For the measurement, Fluke 1760 power quality index analyzers were used. These are class A analyzers and have the appropriate parameters for measuring the aforementioned quantities, according to [11]. The first of the measurement points was selected at the mine's substation, directly behind the 110 kV/6 kV transformer. Thanks to this location, it was possible to determine the parameters of the quality of power supply at the main connection point of the mine.

The second measurement point was located in the medium-voltage 6 kV switchgear in the field from which the winding machine switchgear is supplied. This made it possible to observe the "behavior" of the hoisting machine during the entire

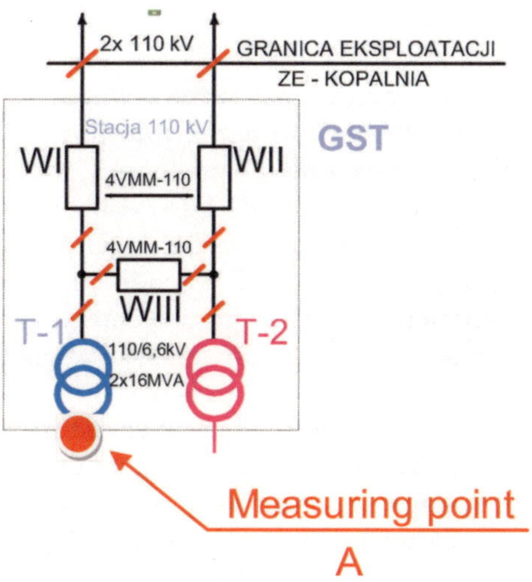

Figure 14.
The first of the measurement points. 110 kV/6 kV substation.

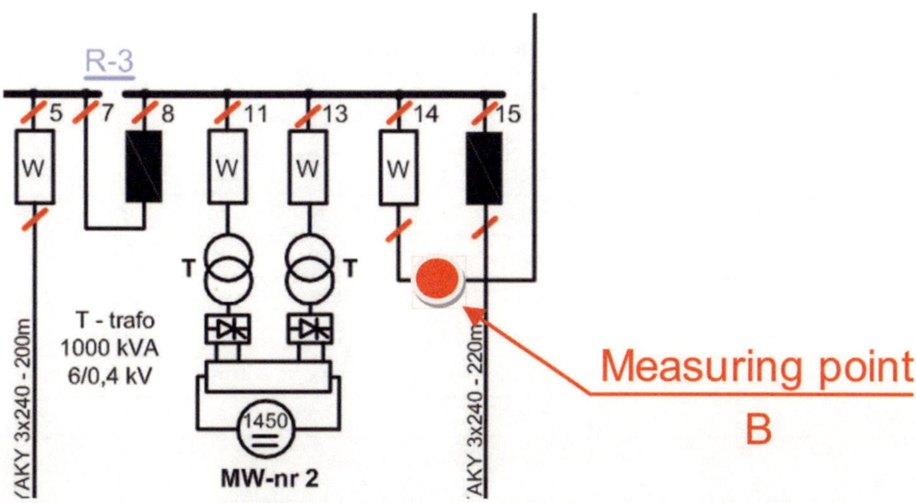

Figure 15.
The second measuring point. Power supply of the winding machine.

duty cycle. No other mine consumers are supplied with this switchgear, so there is no danger of introducing interference from other mine equipment.

From the point of view of the mine's overall power infrastructure, it was important to determine whether the upgraded system has increased the level of introduced disturbances of higher harmonics or voltage variations. In addition to comparing the level of individual voltage harmonics, the level of THD was also analyzed, as well as the Plt coefficient and voltage variations. Total Harmonic Distortion (THD) is defined as the ratio of the RMS value of the sum of all harmonic components (up to the harmonic order 40 in Poland) to the RMS value of the fundamental harmonic [6, 7]:

$$THD_U = \sqrt{\sum_{h=2}^{40} \left[\frac{U_h}{U_1}\right]^2} \tag{12}$$

where

U_h—RMS value of the h-order harmonic of the voltage,

U_1—RMS value of the fundamental harmonic voltage.

The analysis of the system's impact on the power grid included the determination of the levels of individual harmonics and the THD_U voltage distortion factor. **Table 5** summarizes the voltage THD_U for three phases before and after the upgrade. The permissible level specified in [10] was not exceeded in both measurements. However, note an increase in the coefficient after the upgrade compared to the value when the previous system was in operation. This is an increase of about 50%. Therefore from the point of view of the regulations, the system is working properly, but the occurrence of negative phenomena associated with the increase in the level of harmonics in the system cannot be omitted. **Table 5** shows the values of individual harmonics as well. A comparison of the voltage waveforms supplying the winding machine before and after moderrnization is shown in **Figures 16** and **17**. **Figures 18** and **19** show the harmonic spectrum of voltage for the same cases.

Table 5 lists the values of the characteristic harmonics of this system, i.e., 11, 13, 23, 25. The presence of harmonics 5 and 7 may indicate the presence of other

Size normalized (harm. row)	**Values Permissible [%]**	**CP95 [%] Before modernization**			**CP95 [%] After modernization**		
	According to [3]	**UL1**	**UL2**	**UL3**	**UL1**	**UL2**	**UL3**
THDU	8	2.21	2.25	2.18	3.42	3.45	3.48
3	5	0.21	0.25	0.18	0.50	0.26	0.48
5	6	0.30	0.29	0.25	0.46	0.27	0.26
7	5	0.57	0.63	0.60	0.48	0.44	0.43
11	3.5	1.26	1.44	1.37	1.47	1.53	1.55
13	3	1.16	0.87	1.05	1.06	1.04	1.08
23	1.5	0.62	0.74	0.68	0.95	0.94	1.00
25	1.5	0.67	0.57	0.57	0.85	0.84	0.89

Table 5.
THD coefficients and characteristic harmonic levels before and after modernization.

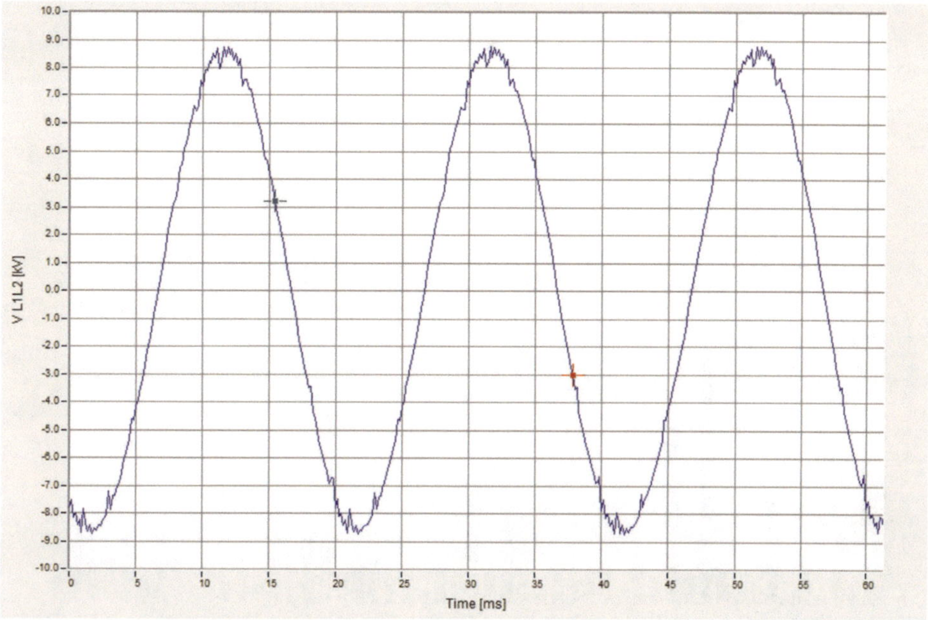

Figure 16.
The course of the voltage supplying the winding machine before the upgrade.

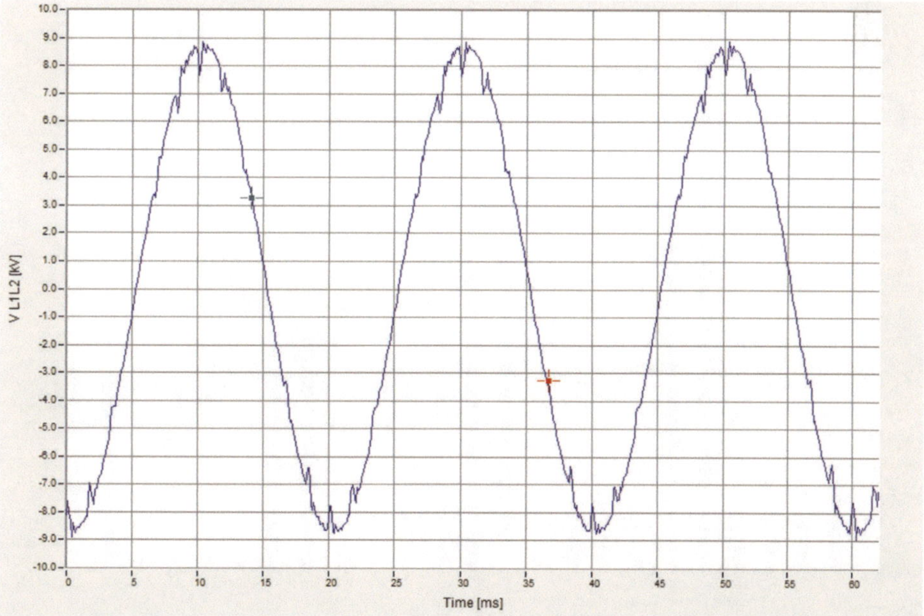

Figure 17.
The waveform of the voltage supplying the winding machine after the modernization.

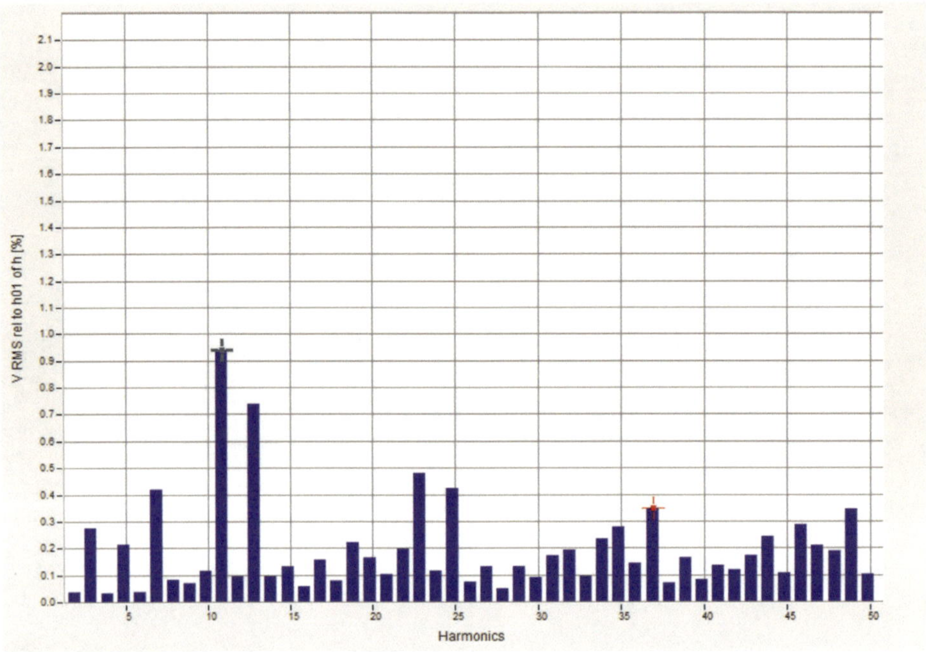

Figure 18.
The harmonic spectrum of the voltage supplying the hoisting machine before the modernization.

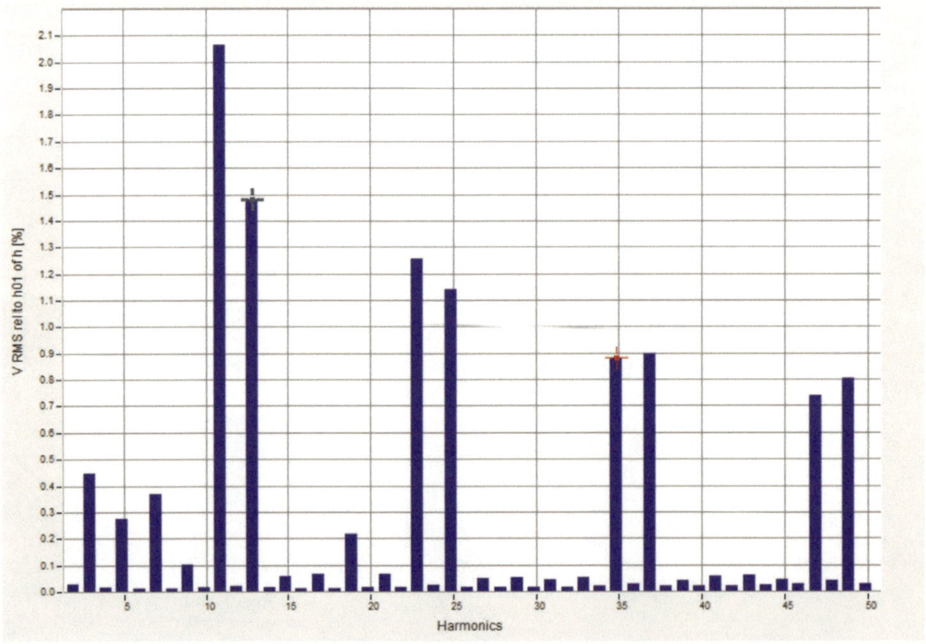

Figure 19.
The harmonic spectrum of the voltage supplying the hoisting machine after the modernization.

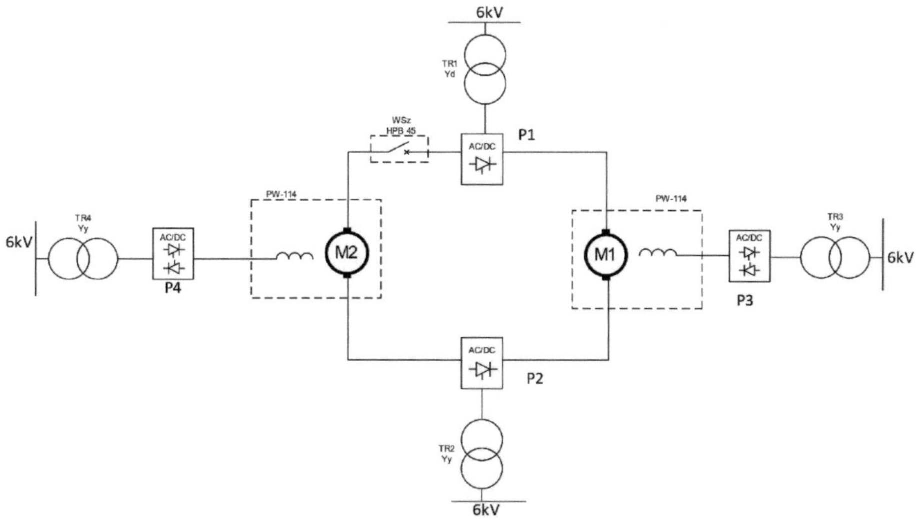

Figure 20.
The diagram of a twin-engine winding machine operating in the Pung system.

disturbances (such as voltage changes) or may be the result of irregularities in the control system of the converters. These harmonics were also present before the system was upgraded. Analyzing the values in **Table 5**, it can be assumed that in the system after the modernization, the cause of these harmonics was partially eliminated, as their level decreased by about 25%. However, an increase in the characteristic harmonics of the system was noted and analyzed.

5.1.3 The modernization number 2

The modernization of the hoisting machine has been carried out at one of the coal mines. The hoisting machine is located on the tower. It has operated in the Pung system since modernization. The block diagram of the system is shown in **Figure 20**.

The hoisting machine consists of two reciprocating motors (M1 and M2) running on a common shaft on which the Koepe wheel is mounted. The main circuit of the hoisting machine consists of series-connected armature motors and DC sides of converters (P1 and P2). This configuration eliminates the interaction of the motors on the mechanical side. The excitations of both motors are supplied from bidirectional bridge rectifiers (P3 and P4). The system can be controlled sequentially or jointly. In the first phase of machine operation, the system was configured to operate with sequence control. This caused interference at other points in the mine's power grid due to the operation of the hoisting machine (**Figure 21**).

The supply voltage waveforms shown in **Figure 22** are distorted. Commutation kinks are clearly visible, occurring at moments when subsequent semiconductor elements take over conduction. The shape of the circuit's supply current significantly deviates from a sinusoidal waveform. The supply voltage with the shape shown in **Figure 22** may pose a threat to the correct operation of the converter due to the difficult synchronization of switching of semiconductor elements.

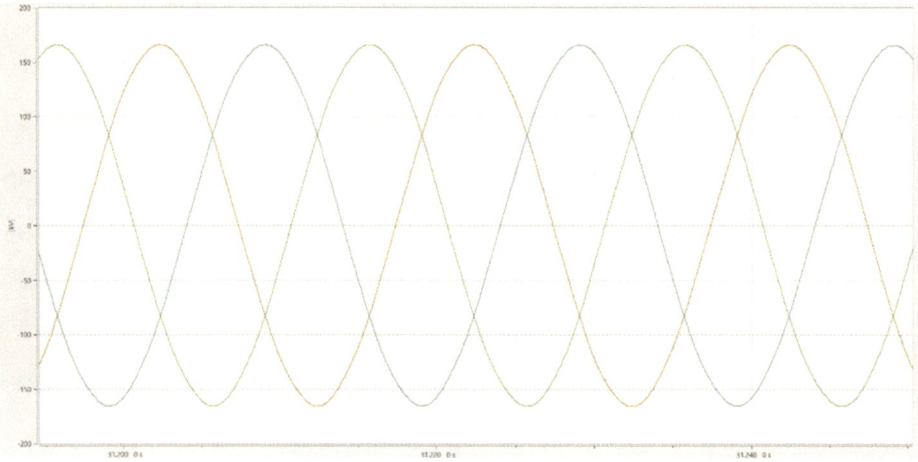

Figure 21.
The power supply voltage waveforms measured at 110 kV substation.

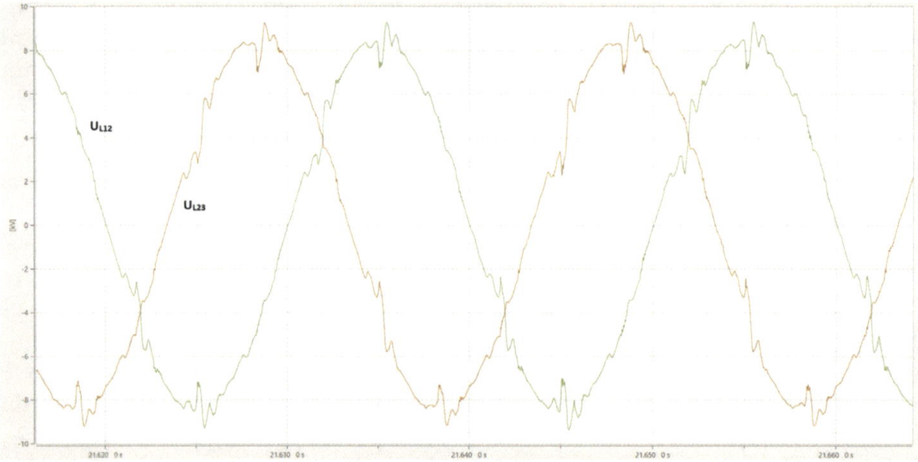

Figure 22.
The course of voltages supplying the winding machine.

The spectrum of harmonics indicates the operation of a 6-pulse system. The largest harmonic values in relation to the fundamental harmonic are reached by harmonics fifth and seventh (**Figure 23**). The presented waveforms of phase-to-phase voltages occurring at the mine's power supply point speak for the absence of significant distortion due to equipment operation (**Figure 21**). This means that distortions negatively affect equipment operating inside the network at the medium-voltage level of 6 kV. The distance between devices can also negatively affect the amplification of a certain type of disturbance. In this case, the distances between devices are several kilometers (**Figure 24**).

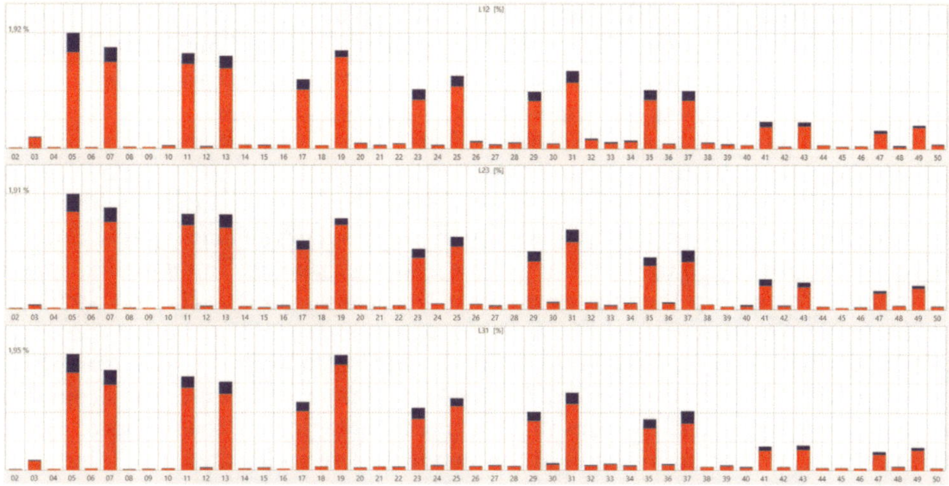

Figure 23.
Harmonic spectrum of the phase voltages.

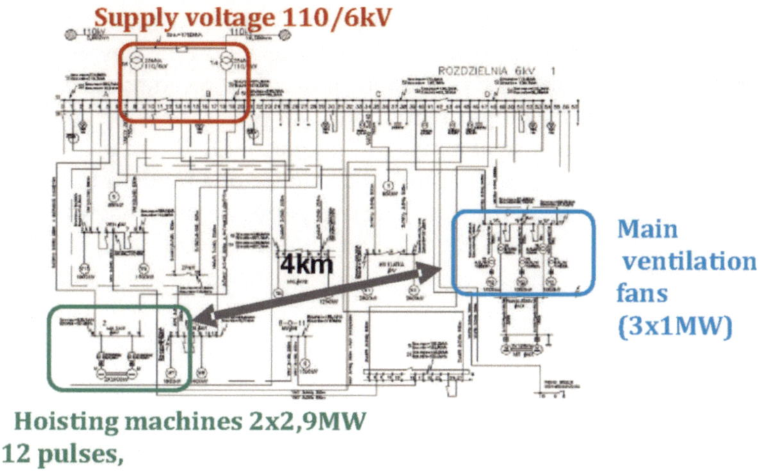

Figure 24.
The mutual location of equipment in the power grid.

6. Conclusions

Modernization of mining shaft hoists is among the largest investments in the structures of mining plants. Therefore, it is important to prepare the entire project properly, with special attention to the power supply side of the hoisting machine. As an electrical device with a high degree of complexity, the hoisting machine requires a power supply with appropriate parameters. On the other hand, it is one of the largest devices in the mine network that is a source of disturbance, and is therefore a real threat to other consumers of the grid and safety systems.

In the case presented in Section 4.1, the preliminary analysis became the benchmark for determining the level of impact of the newly created circuit. That biggest concern was the change in power circuit configuration. In Poland, more than 90% of hoisting machines using a DC reciprocating motor operate in a system with reversion through the excitation circuit. There was concern that the use of high-power four-quadrant converters in the armature circuit would cause interference in the mine network to increase beyond acceptable values. The tests carried out did not confirm this thesis. Therefore, it should be concluded that the use of a system with reversion through the main (armature) circuit is justified from a technical point of view because it does not cause an increase in the levels of higher voltage harmonics above the limits. The advantage of this system is the increase in the speed of the hoisting machine, which can be reflected in the efficiency of the hoisting machine. This situation appears due to the differences in the time constants of the armature and excitation circuits.

The conclusions that emerged from the measurements could not be considered as a basis for discontinuing the monitoring of power quality indicators in the network at this mine. It should be borne in mind that the level of interference increased compared to the previous system. This means that the phenomena that are the results of the increase in the level of interference in this network were anticipated. In a mine network, this is important as long as it can affect protection systems, which is a potential danger to the crew. Shortly after the completion of this stage of measurements, a ferroresonance phenomenon occurred in the network, which contributed to significant property damage.

Currently, technical possibilities are being searched for to build hoisting machines in a way that significantly reduces their negative impact on the power grid. One such solution is the use of slow-speed synchronous motors powered by intermediate frequency converters with active input rectifiers.

Another possibility is the use of active higher harmonic filters in already operating winding machines with DC helical motors. The choice of a method to reduce the negative impact of the drive on the power grid depends on many factors and should be analyzed in each case.

Author details

Tomasz Siostrzonek
Faculty of Electrical Engineering, Automatics, Computer Science and Biomedical Engineering, Department of Power Electronics and Automation of Energy Transformation Systems, AGH University of Science and Technology, Krakow, Poland

*Address all correspondence to: tsios@agh.edu.pl

References

[1] Dugan RC, McGranaghan MF, Santoso S, Wayne BH. Electrical Power Systems Quality. Mc-Graw-Hill; 2004. Chapter 5 and 6

[2] Shepherd W, Zakikhani P. Energy Flow and Power Factor in Non-sinusoidal Circuits. New York: Cambridge University Press

[3] Karve S. Harmonics – Active harmonic conditioners. Leonardo power quality application guide – Part 3.3.3. 2001

[4] Hanzelka Z, Bien A. Harmonics – Interharmonics. Leonardo power quality application guide –Part 3.3.1. 2004

[5] Wilsun X, Xian L, Yilu L. An investigation on the validity of power-direction method for harmonic source determination. IEEE Transactions on Power Delivery. 2003;**18**(1):214-219

[6] Hanzelka Z. Jakość Dostawy Energii Elektrycznej. Wydawnictwa AGH: Zaburzenia wartości skutecznej napięcia; 2013

[7] Baggini A. Handbook of Power Quality. England: John Wiley & Sons, Ltd.; 2008. Chapter 4 and 7

[8] Energoelektronika PS. Układy o komutacji sieciowej oraz komutacji twardej. Uczelniane Wydawnictwa Naukowo-Dydaktyczne AGH; 2006

[9] Mining State Authority. Available from: https://www.wug.gov.pl/bhp/nadzorowane_zaklady [Accessed: June 20, 2023]

[10] Regulation of the Minister of Energy of 23 November 2016 on detailed requirements for the conduct of underground mining operations

[11] Regulation of the Minister of Climate and Environment of March 22, 2023 on detailed conditions for the operation of the electric power system

[12] Zhang X, Yan Z. Energy quality: A definitione. IEEE Open Access Journal of Power and Energy. 2020;**7**:430-440

[13] Bebikhov YV, Egorov AN, Semenov AS. How higher harmonics affect the electrical facilities in mining power systems. In: International Conference on Industrial Engineering, Applications and Manufacturing (ICIEAM). Sochi, Russia; 2020

[14] Wiechmann EP, Aqueveque P, Munoz L, Henriquez JA. Energy quality and efficiency of an open pit mine distribution system: An improvement. IEEE Industry Application Society Annual Meeting. Vancouver, Canada; 2014

[15] Agarwal A, Kumar S, Ali S. A research review of power quality problems in electrical power system. MIT International Journal of Electrical and Instrumentation Engineering. 2012;**2**(2):88-93

[16] Chapman SJ. Electric Machinery Fundamentals. New York: McGraw-Hill; 2005. Chapter 8

Section 2

Energy Security in the Era of Climate Change and Growing Uncertainties for Resilience in Sustainable Energy Development

Chapter 7

Perspective Chapter: The Regime Matters – A Multidisciplinary Perspective on Energy Security in the Era of Climate Change and Growing Uncertainties for Resilience in Sustainable Energy Development

Smart Edward Amanfo and Joseph John Puthenkalam

Abstract

This chapter provides a comprehensive exploration of energy security, beginning with various definitions and their positioning within the international political economy of energy. It examines energy security from multiple angles, including perspectives of energy-importing and exporting nations, transit regions, militarization, energy shocks, demographic shifts, and corruption. The chapter highlights the challenges, risks, and vulnerabilities inherent in energy security and underscores its intricate interconnections. It concludes by advocating for the integration of resilience thinking into energy security policies due to growing uncertainties in social, economic, and ecological systems, compounded by climate change. These factors significantly shape the context in which energy security strategies are devised and implemented.

Keywords: energy security, climate change, uncertainty regime, energy supply-demand security, resilience sustainable energy development

1. Introduction

Exosomatic energy is putative of global economic growth and development [1] and remains at the center of contemporary socio-economic development and environmental sustainability policy discourses. As a result, the security concern of exosomatic energy is one of the contemporary sustainable development priorities across the globe. Giving the above reason, every national economy endeavors to ensure a sufficient degree of energy security and formulate, strategize, and execute energy security policies in accordance with individual economies' requirements and unique circumstances [2]. The notion of energy security is multifaceted and dynamic–it revolves around the understanding and analysis of the factors, dynamics, and policies that contribute to ensuring a stable, reliable, and sustainable energy supply. Given the centrality of energy and entire socio-technical systems – its security discourses encompass various dimensions, including, but not limited to, political, geopolitical, economic, social, and socio-technological systems components, and aims to mitigate risks and vulnerabilities associated with energy systems across time and space. In line with the above reasoning, the authors in [3] have suggested that the concept of "*energy security*" is more akin to an abstraction than a well-defined policy or term, making it challenging to describe. Furthermore, whether one reason from the positivist or normativist (or a combination of the sort) perspectives, the conceptual and practical meaning of energy security can vary depending on different perspectives, such as institutional, national, or personal views [4]. Again, from the international political economy of energy perspective, securitization of energy system is a construct of four main frames – namely: the "dominant worldviews", "prioritized component of energy security", "energy security for whom?", and "underlying values and goals" – and are influenced by the ruling social, economic, political or environmental ideologies (see **Table 1**). Humans are evolutionarily wired to be *security-thinking burdened* species – thus, how to secure energy is human existential. For example, a view exists that human conception of energy security can be traced to the prehistoric era when humans first discovered and began controlling fire [6] as an exosomatic energy. However, the beginning of academic discourses on energy security might have emerged in the 1960s [7], probably as one of the post-Second World War development thinking. But, the 1970s oil crisis fairly marked the commencement of energy securitization and measurement efforts through policy rationalization and academic research [8, 9], especially among the OECD (Organization for Economic Cooperation and Development) economies. A year after the 1973 oil crisis – 1974, the International Energy Agency (IEA) was established as an intergovernmental organization to promote energy supply security and foster economic growth – a direct response to the oil crisis that unfolded during the years 1973–1974 [8]. According to an account, as part of the IEA's oil crisis management frameworks, the OECD (Organization for Economic Cooperation and Development) member countries must hold oil stocks

Frame Agents	Dominant worldviews	Prioritized components of energy security	Energy security for whom?	Underlying values and goals
Market liberalists	Technological optimism, free market libertarianism	Economic affordability	Economy	Welfare, freedom
Neo-mercantilists	Defense of national security	Geopolitical availability	State	Political independence and territorial integrity
Environmentalists	Environmental preservationism, conscientious consumption	Environmental sustainability	Earth	Respect for nature
Social greens	Justice, neo-Marxism	Social acceptability	Society	Equity, justice

Source: Adopted from ([5], p. 93).

Table 1.
Frames and worldviews on the international political economy of energy.

equivalent to 90 days of imports [8], as part of energy security measures. One can fairly argue that in years predating the first oil crisis in the 1970s, the securitization of energy systems was not a prominent policy issue [9]. With the turn of events, however, the popularity of social, economic, political, and geopolitical impacts of the 1973 oil crisis became a household concern, especially in the developed economies that largely relied on the Gulf States region for oil. As "loosely" emphasized by ([10], p. 111), "even casual newspaper readers have become aware that there are links between energy, security and foreign policy". Notable early works from a political point of view, like Willrich's article on international energy issues and options, provided comprehensive analyses of the international energy sector and its implications on a national and global scale [11]. Willrich distinguished between "*security of supply*" and "*security of demand*" and recognized that energy issues varied for different stakeholders, be they oil-importing or producing nations [p. 746]. Throughout the latter part of the 20th century, energy security studies primarily operated within a political economy framework with a focus on a reasonably stable supply of oil [12], with little or no recourse to environmental sustainability dimensions. Researchers and policymakers prioritized diversification, uninterrupted flow, and affordability of energy supply. Such earlier rationalization of the energy security concept renders it as "security of supply" focused, and as a central element in defining and operationalizing energy security. If we turn to the use of the term "security" as constantly applied by security researchers and scholars, *security* collocates the "non-existent of threats". We are however further burdened with the word "threat" – as spatial, time, culture, values and perspective dependent – implying the concept of security co-evolves social, economic, environmental, geopolitical, and the like factors. For instance, during the First World War (WW I) and Second World War (WW II), access to refined, efficient energy fuels was a topic profile geopolitical and national security strategy reflecting the Neomercantilists worldview of seeing energy security as a defense of nationality security (see **Table 1**).

Today, as the body and internal structure of knowledge about the "indispensability of energy systems" – as humans cannot survive without working – and no work could be done without useful energy services, energy security scholarships have evolved into an interdisciplinary field. This is largely due to factors such as climate change, globalization, and the uncertain future of fossil fuels. These developments have introduced new dimensions to the concept, including sustainability, resilience, energy efficiency, greenhouse gas emissions mitigation, accessibility of energy services (addressing energy poverty and its multidimensionality), and more. As a result, energy security has become intricately intertwined with other environmental, social, political, and security issues. To capture the multidimensional nature of energy security, systematic literature reviews [13, 14] and international research [15] have explored diverse perceptions and indicators of the concept. In particular, [15] went further with the dominant focus on finding a converging definition for the concept of energy security by identifying 16 distinct dimensions of energy security – including affordability of energy services, equitable access to energy, and energy efficiency. Sovacool et al. [15] go beyond traditional definitions, including dimensions like energy education and ensuring transparency in their research. Through questionnaire administration, these authors rigorously demonstrate that people's perception of energy security varies and covariates education, age, gender, and culture value systems. Sovacool [16] further expands on this idea by introducing the concept of "cultures" within the energy sector, suggesting that different perceptions can be explained by the cultural context to which the participants belong. Additionally, Sovacool [16] identifies at least five different energy cultures, including national, economic, political, professional, and epistemic cultures. By doing so, the author highlights the multifaceted nature of energy security and emphasizes how various subjects perceive the concept differently based on their own preferences and backgrounds.

Another attempt to conceptualize energy security is in the work of [17]. These researchers focus on the core question of "*what to protect?*". Accordingly, they define energy security as the "low vulnerability of vital energy systems". The energy systems' "vulnerability" may result from exposure to risks – being they natural or due human factors (or a combination human and natural factors). In the context of low-income economies that faced with weak, and inequitable critical energy infrastructure, but are faced with extreme climatological events due to climate change and variability [18], building resilience capacity is indispensable to maintaining security. The authors examine vulnerabilities across different energy systems, including energy infrastructure, energy services, and renewable energy sources. While this introduces some senses of universalism in our understanding of energy security, the authors acknowledge that energy security remains an abstract concept that heavily depends on the perspectives, actions and values of the actors involved. In retrospect, consensus exists that the meaning of a concept, in general terms, changes over time. Citing the concept of "democracy" as an example, ([19], p. 18) suggests that "Democracy 2000 years ago had a different meaning than democracy today". By a way of cross-disciplinary reasoning, we can claim that the concept of energy security is continuously evolving, with its scope continually expanding and lacking a universally accepted definition. That notwithstanding, promoting sustainable development relies significantly on ensuring energy security, which has become a fundamental pillar in energy policies and frameworks worldwide. According to Chris Rupped (n.d.), as cited in [20], the period 1985 to 2003 marked the era of energy security, and from 2004 onwards, marked the period of "*energy insecurity*" [p. 172]. With the growing uncertainties and vulnerabilities in social, economic, political, geopolitical, and environmental systems, including emerging infectious diseases, that follow the patterns of global warming and climate [21], the context in the notion of energy security, relative to the three major policy and academic questions: (a) "security for whom", (b) "security for which values", and (c) "security from which threats", is likely to maintain an enduring energy systems research and policy concern far into the twenty-first century.

1.1 Dynamics and perspectives of energy security thinking

Energy security's interpretation varies in academic literature, among different stakeholders, and across energy-importing/exporting nations, organizations, and military. The broader understanding is shaped by strategies for safeguarding it. This multifaceted concept spans diverse dimensions, explored next to understand varied actor perspectives.

2. Net-importing (relatively energy resource poor) economies energy security

Energy resources exhibit varying geographical distributions and availability [22, 23]. In countries reliant on energy imports or those lacking ample resources, ensuring energy security centers on maintaining a consistent, reliable energy supply.

This involves addressing citizens' energy needs and sustaining crucial sectors like transportation, heating, and electricity [24]. According to [11], energy security pertains to guaranteeing adequate energy supplies to sustain the national economy at a standard level.

Hence, economies reliant on energy imports prioritize enacting effective domestic and foreign policies to ensure a stable energy supply. These actions may encompass diversifying energy sources, enhancing efficiency, and promoting new technologies [24–26]. These strategies aim to reduce dependence risks build robust energy infrastructure, and led to the formation of the International Energy Agency (IEA) in 1974 – a global organization of 29 economies focused on energy security. Established within the framework of the OECD, the IEA characterizes energy security as "as the uninterrupted availability of energy sources at an affordable price"[1], which is consistent with three out of the popular "the 4 A's of energy security"[2] – *availability*, *accessibility*, and *affordability* dimensions of energy security suggested in the literature [17]. Consequently, the IEA's energy security definition aligns with that of the European Union (EU). As the title of this chapter seeks to describe, regime thinking matters in the attempt to conceptualize energy security and determine the most optimal strategic responses to deal with short-term supply or demand-side shocks. As a result, the IEA has recently added the imperatives of climate change, weather, and digital resilience to energy security rationalization, including environmental sustainability.

Europe does not only represent a significant energy import-dependent region but has also traditionally relied on individual member states and their national initiatives for energy security strategies. However, in recent decades, the European Union (EU), as an intergovernmental organization, has characterized its energy policy in three main thematic areas: *energy security*, *competitiveness*, and *sustainability* ([28], p. 6). The Energy Security Strategy of the European Union addresses both immediate and future energy security measures. In the short term, the strategy primarily emphasizes enhancing resilience against energy supply disruptions through diversification of energy supply sources. Meanwhile, the long-term measures aim to decrease dependence on external energy sources. To ensure the ability to overcome supply disruptions, the strategy proposes various actions, such as coordinating risk assessments, establishing reserves, safeguarding critical infrastructure (particularly focusing on cybersecurity), fostering an integrated internal market (including the construction of vital inter-connectors among member states), strengthening cooperation with new suppliers, and developing new energy transit routes. An example of such a route is the Southern Gas Corridor [29]. Given European Union Member States over reliance on Russia for energy, especially gas, the resilient of the region's energy systems has to be tested in the face of the ongoing Russian-Ukraine war and the energy sector impact of the COVID-19 pandemic. A seeming progress in line with the above, however, is that, European Union Member States reportedly managed to reduce their dependency on Russia in their natural gas imports by 40% in 2022, compared to 2019 ([30], p. 64).

In the realm of global energy geopolitics, the United States stands uniquely; it is faced with challenges and opportunities inherent in both energy-importing and exporting economies. Given this complexity, the Energy Information Administration (EIA) [31]

---

[1] The reader may read details from https://www.iea.org/areas-of-work/energy-security.

[2] The Asia Pacific Energy Research Centre (APERC) provides a contemporary dimensions of energy security – Availability, Acceptability, Accessibility, and Affordability (The 4 A's of Energy Security [27]).

argues that the United States holds the distinction of being both the largest energy importer and the largest energy producer and consumer globally. However, there is a growing consensus that the shale revolution is transforming the United States into a net exporter of oil and gas. Meanwhile, the US Energy Information Administration further projects that the country will become a net energy exporter by 2022. However, it seems the USA became a net exporter of petroleum two years ahead of the 2022 projection by the EIA, as the same source indicates that "In 2020, the United States became a net exporter of petroleum for the first time since at least 1949". For instance, in 2022, the US exports of petroleum were estimated as 9.58 million barrels per day (b/d) as against 8.32 million b/d in the same year under review. Making America's annual net petroleum exports 1.26 million b/d in 2022.

Traditionally, energy security in the United States has primarily focused on achieving what is known as "energy independence", particularly concerning oil. However, as observed by a prominent energy policy scholar [16], neither the Energy Independence and Security Act of 2007 nor the Food and Energy Security Act of 2007 (both of which were signed into law by President Bush) provide a clear definition of energy security. An exception of the recent official document that seeks to address the notion of energy security is a paper from the Executive Office of the President of the United States ([32], p. 11). It establishes that "*energy security is used to mean different things in different contexts and broadly covers energy supply availability, reliability, affordability, and geopolitical considerations*". This aligns with the definitions provided by the United Nations (UN) and the International Energy Agency (IEA). It is widely acknowledged that the shale revolution is playing a significant role in transforming the United States into a net exporter of oil and gas.

When it comes to energy security perspectives from net-importing economies, Japan stands out as a substantial "elephant in the room" – with its significant footprint on the global energy landscape. Due to the country's disadvantage in terms of the geography of strategic energy resources, the country heavily relies on imported energy commodities. For example, according to the World Nuclear Association, Japan imports approximately 90% of its energy requirements.[3] In 2017, the Ministry of Economy, Trade and Industry estimated Japan's primary energy self-sufficiency as 9.6% and ranked 34th among the OECD countries [33]. Consequently, supply-side security has always been both domestic and geopolitical concern. Thus, energy dependence – the extent to which internal/external disruption to energy availability would negate the welfare of citizens and energy security (already defined) has always been an integral part of Japan's national security strategies through times of major energy crises. As a reference period, the oil embargoes of the 1970s substantially unearthed and exposed Japan to the risks associated with relying on oil from the Persian Gulf States. As a result, by 1985, nuclear plants constituted about 23.8% of total electricity generation and remained steady until 2010 – with a share of about 25.7%. Following the nuclear accident at Fukushima in March 2011, nuclear plants share of electricity generation reduced to approximately 14.7% and to .0.00% in 2014 due to the subsequent closure of all Japanese nuclear reactors. Japan's "pride" is a function of a stable or secured energy system. Thus, the 2011 Daiichi Nuclear plant meltdown led to a sharp increase in liquefied natural gas (LPG) imports. Some analysts argued that the Fukushima Nuclear accident in 2011 affected both the supply and

[3] Details available from https://www.world-nuclear.org/information-library/country-profiles/countries-g-n/japan-nuclear-power.aspx. Seen 6 August 2023.

demand sides energy security globally – as Japan's demand for fossil fuels surged, raised the prices of liquefied natural gas (LNG) and uranium [34] – affecting availability, affordability, sustainability and accessibility dimensions of energy security as mentioned in the reviewed literature. Specifically, using 2008 as a reference year, Japan's consumption of natural increased from 88 billion cubic meters (bcm) to 114 bcm in 2011, 119 bcm in 2014, and steadily declined to the 2009 level of 93 bcm, and 92 bcm in 2022.[4]

Surprisingly, the "Strategic Energy Plan 2018", and the "Japan 2021 Energy Policy Review" [33, 35], pivotal energy policy documents in Japan, overtly highlight the escalating apprehension regarding energy insecurity. It delineates a comprehensive framework for enhancing security, encompassing various measures such as augmenting the availability of fossil fuels both domestically and internationally, hastening energy efficiency initiatives, fostering the growth of renewable energy sources, reconfiguring nuclear policies, opening up energy markets, and formulating strategies for swift emergency responses following the Fukushima incident. Under Tokyo's Strategic Energy Plans, energy security is prioritized. In this policy logic, the Japanese Government's stated goal of achieving energy security, among other things, is to "increase the 'independent development ratio' ... to 40% for oil and natural gas, and to 60% for coal by 2030" ([35], p. 30). Paradoxically, even though this plan emphasizes these security measures, the official elucidation of the energy security concept by the Government of Japan has been omitted.

3. Energy exporters landscapes

When considering energy security, economies abundant in energy resources also share security concerns. In that, socioeconomic development, national security, cyberinfrastructure, etc., are significantly influenced by energy security, just like energy-importing countries. Thus, energy security encompasses both supply-side and demand-side security. Policymakers energy security burden from the landscape of the energy-exporting economies should be defined as the "security of supply". Here, security of demand can be viewed as the guaranteed access to foreign markets for energy exporters. This fairly implies that energy security as a policy concept from the energy-importing and energy-exporting economies are not mutually exclusive. Exporters have to strategize to deal with both such external and internal variables, including but not limited to access to international energy markets, global demand and prices, domestic and geopolitical stability, investments and infrastructure, market diversification, environmental regulations, the rate of energy resource depletion, major environmental treaties and conventions, currency fluctuations and sanctions. In particular, market diversification is imperative to maintain a stable stream of revenue. This is very crucial for the energy-rich-poverty paradox countries, such as Nigeria, Ghana, Angola, Gabon, Chad, and the Republic of the Congo. With specific reference to Africa, both matured and new oil-exporting countries have to strictly adhere to the principle of the Non-Aligned Movement established at the Belgrade Conference in 1961, probably to avoid spillover effects of sanctions as have been applied to the powerful Western allies against countries such as Russia and Iran during geopolitical

---

[4] Analysis based on data from blob:https://yearbook.enerdata.net/bf89d04c-7e1d-49ea-aba9-f9b6e1df4d5e. Seen 6 August 2023.

tensions. Another aspect that can enhance demand security is maintaining a delicate equilibrium between supply and demand. Energy exporters (most resource-rich economies) are interested in maintaining stable demand for their energy products, which is often linked to the interests of the industrialized club economies and the development of non-fossil energy sources in the climate regime.

From the landscape of the energy-exporting countries and the concept of energy security, the role of the Organization of the Petroleum Exporting Countries (OPEC) is worth considering with reference to 1970s. As a general knowledge, the OPEC is an intergovernmental organization consisting of 13 oil-exporting states. While not all major energy suppliers, such as Russia, are members of OPEC, the organization accounted for 36.3% in 2022 [36] as against 42.4% in 2017 [37] of global oil production, making it the largest energy exporter under the period of review. In the 2010 Long-Term Strategy document, OPEC outlined long-term energy policy objectives of ensuring long-term oil revenues by stabilizing fair energy prices and securing global oil demand for producers. The official document acknowledges the importance of understanding future oil demand requirements as a crucial aspect of OPEC's overall energy security concerns. Additionally, the strategy emphasizes security of supply, which, in OPEC's context, refers to ensuring an efficient, economic, and consistent supply of petroleum to consuming nations.

However, it is worth noting that while OPEC primarily focuses on energy security purely from the perspective of energy exporters, it recognizes shared energy security concerns with the rest of the world. OPEC acknowledges that environmental issues like climate change have global implications, necessitating multilateral cooperation among all stakeholders in the energy sector. It is worth mentioning that, based on our current social-technical systems, energy stands as a pivotal resource driving transformative global development, with oil and gas maintaining their crucial roles within this sector. Despite the ongoing shift toward renewables, these traditional sources will remain integral, particularly in developing and emerging economies for the foreseeable future in addressing other dimensions of security, including energy poverty, food and water security, health, sustainable transport, dealing with emerging infectious diseases, climate change and global warming adaptation, etc.

Energy is a key resource for transformational development globally. Oil and gas (fossil fuels) will continue to play a key role in this sector, irrespective of the gradual transition toward renewables, and will continue to do so in most developing and emerging economies in the near future.

4. Militarization and energy security

Militarization – defined as "the cultural, symbolic, and material preparation for war" [38], can have significant implications for energy security, both directly and indirectly. In military parlance, energy security is often defined as operational energy. This technically encompasses the daily energy required for training, transportation, and sustaining military forces and weapons platforms during military operations. Emerging out of this global recognition is that energy security has become an increasingly important strategic aspect of military planning and activities. Multiple forces are propelling this trend, but can be summarily enumerated under such sub-headings as the desire for resource control and geopolitics, concerns about supply disruptions, vulnerability of energy infrastructure network systems, energy transit routes, environmental concerns (global warming and climate change), high defense spending, dependence on energy imports, supply disruptions.

Militarization can lead to the control or occupation of energy-rich regions or key transit routes for energy resources. This can potentially grant a country greater influence over energy markets and supply chains. Geopolitical tensions arising from militarization can disrupt the flow of energy resources, impacting global energy markets and prices. Again, Militarized conflicts in energy-producing regions can disrupt the extraction, production, and transportation of energy resources. Attacks on energy infrastructure, such as oil fields, refineries, pipelines, and ports, can lead to supply disruptions and price spikes. This instability can affect energy-importing nations' access to vital resources. Based on history and contemporary wartime experiences, energy infrastructure systems are often a target in conflicts due to their economic and strategic importance. Militarization can potentially heighten security concerns for energy facilities, making them more vulnerable to attacks, sabotage, and cyber threats. This can undermine the reliability of energy supply chains. From scarcity and allocation reasoning, military buildups and maintenance of a large military force require significant resources, including energy. A country with a highly militarized stance might divert resources away from the development of renewable energy sources and energy efficiency measures, thereby hindering its transition to a more sustainable energy mix.

Essential to the above is a "collective defense" organization, like the North Atlantic Treaty Organization (NATO). Until the recent decades of heightened environmental and ecological security concerns, NATO's foremost focus has been to function as a collective and regional security organization [39], particularly in defense against the Soviet Union following the aftermath of the Second World War. Indisputably, energy security is the nucleus of the collective security of its members. Keshk [40] has eloquently stated that *"energy security is one of red lines crossing which by any party is not allowed by NATO"* ([40], p. 10). Accordingly, disruptions in energy supply (particularly fossil fuels) can have significant implications for the security of NATO allies and partners, potentially impacting the effectiveness of NATO's military operations. Typically, energy security policy would traditionally fall within the purview of individual sovereign nations. Nevertheless, NATO allies are committed to engaging in discussions regarding energy security, encompassing its broader dimensions, and collaborating to bolster NATO's ability to make meaningful contributions in areas where energy has the potential to yield collective or regional benefits across social, political, economic, and even ecological domains.

For instance, in their Strategic Concept for the Defense and Security document, the Members of the NATO explicitly state that

> *Key environmental and resource constraints, including health risks, climate change, water scarcity and increasing energy needs will further shape the future security environment in areas of concern to NATO and have the potential to significantly affect NATO planning and operations ([41], p. 15). [Emphasis added].*

In essence, the process of militarization can lead to extensive outcomes on energy security. This encompasses impacts on resource access, supply chains, infrastructure, and the potential to shift toward more sustainable energy sources. Moreover, tense geopolitical relations and conflicts possess the capacity to interrupt energy markets, sway investment choices, and impede the overall progress toward establishing a dependable and environmentally-friendly energy future.

5. Energy transit economies landscape of energy security

Conventionally, energy particularly commercialized fossil fuels, entails intricate and interlinked value chains encompassing various stages such as exploration, extraction, conversion, transportation, and distribution to potential end-user sub-sectors. Thus, energy resources can be categorically delineated as the primary basis for dividing the global landscape into three distinct "energy worlds": nations endowed with abundant energy resources, nations primarily characterized as energy consumers, and nations serving as intermediaries for energy transit [42]. Another context is that in 2022, about 82% of the total primary energy consumption globally came from fossils – with oil, gas, and coal, respectively, accounting for 31.6%, 23.5, and 26.7%.[5] Meanwhile, globally, oil and natural gas – constituting about 55.1% of total primary energy consumed in 2022, were (as has been historically) exclusively transported through sea routes. Consequently, we reasonably argue that the attainment of energy security – whether collectively or individual sovereign nation perspectives, hinges significantly upon the primacy of energy diplomacy and cooperative efforts among both importing and exporting nations, *ceteris paribus*. This places the world's major *chokepoints* – defined as "narrow channels along widely used global sea routes" [44]. According to [44], global chokepoints for maritime transit of crude products (particularly oil and gas) are a critical component of global energy security. For instance, in 2015, the Energy Information Administration estimated that 61% of global petroleum and other forms of liquids were transported through maritime routes. Although the statistics are somewhat outdated, their validity might still hold true, given the absence of recent evidence indicating any changes in the geographical and associated energy geopolitical fundamentals in these regions.

The EIA has named seven chokepoints: the Strait of Hormuz, Strait of Malacca, Suez Canal, Bab el-Mandeb, Cape of Hope, Danish Strait, Turkish Straits, and Panama Canal, whose security determine the faith of global energy systems' security (See **Figure 1**).

Just for reference purpose, the Suez Canal and the SUMED Pipeline serve as strategic conduits for the transportation of Persian Gulf crude oil, petroleum products, and liquefied natural gas (LNG) shipments to Europe and North America. Situated within Egypt, the Suez Canal effectively links the Red Sea with the Mediterranean Sea, assuming a pivotal role as a critical chokepoint due to the substantial quantities of energy commodities that traverse this passage. In 2018, for instance, the northbound petroleum exports from the Persian Gulf States – Iraq, Iran, and Saudi Arabia to Europe and North America, 85% were reportedly transported via the Suez Canal.[6] In the realm of energy security policy discourse, the pivotal point is that the essential elements of energy security, denoted as the 4 A's, rely fundamentally on the stability of key energy transit routes and chokepoints, as previously indicated. Some energy risk analysts have strongly reiterated that blocking a chokepoint, even temporarily, can lead to substantial increases in total energy costs and world energy prices – negating the affordability dimension of energy security. In addition, chokepoints also leave oil tankers vulnerable to theft from pirates, terrorist attacks, political unrest in the form of wars or hostilities, and shipping accidents that can lead to disastrous oil spills [45].

---

[5] Computation based on data from https://ourworldindata.org/energy-mix [43]

[6] Details available from https://www.eia.gov/todayinenergy/detail.php?id=40152. Seen 25 July, 2023.

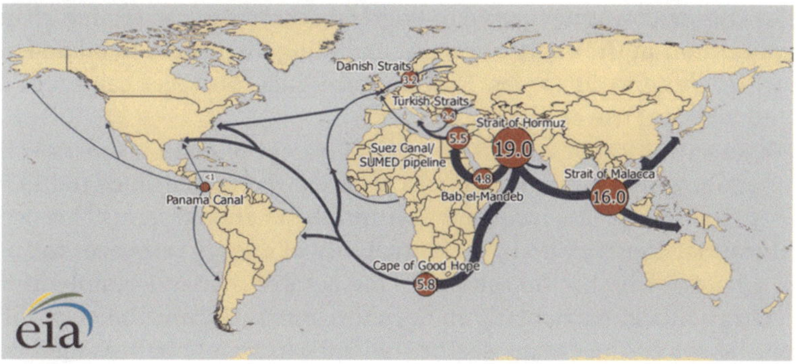

Figure 1.
World oil transit chokepoints: Source: Adapted from [44].

While sustainable development goals and other international initiatives have emphasized the significance of addressing energy security, which involves the intricate interplay between poverty reduction and the mitigation and adaptation of climate change, the intricate nature of energy interdependence facilitated by geographical attributes such as energy transit routes and chokepoints has not been systematically deliberated upon. In order to effectively attain the energy objective outlined in the sustainable development goals, these deliberations must assume a prominent position within the framework of the United Nations system. Valuable insights can be drawn from the profound impacts of events like the COVID-19 pandemic and the Russia-Ukraine war, which have starkly exposed vulnerabilities within global energy systems, particularly the disproportionate repercussions on economies with limited financial resources. Thus, in the regime of high uncertainties like climate change, geopolitical tension, pandemics, etc.; low-income economies, in particular, must raise energy security policy questions duly relevant to the physical, economic, markets, social, and geopolitical imperatives of energy system.

6. Energy shocks (crises)

Energy crises or shocks and human progress constitute co-evolutionary phenomena within social-ecological systems. While the oil shocks of the 1970s predominantly hold the status of a reference point in energy economics and historical literature, an alternative proposition posits that the initial documented energy crisis took place around 1600 – known as the "Firewood crisis in England" ([46], p. 417). Thus, throughout history, energy crises have arisen due to various factors such as energy shortages (resource depletion), conflicts, market manipulations, rent-seeking behaviors, and government interventions. Fiscal interventions – such as tax increases, the nationalization of energy companies, and regulatory measures in the energy sector can disrupt the balance of supply and demand, leading to an intermittent or prolonged crises. Additionally, protests, government embargoes, over-consumption, aging infrastructure, disruptions at oil refineries and port facilities, and increased energy consumption during harsh summers and winters can also contribute to emergencies.

Minor interruptions in energy supplies, especially in countries with energy infrastructure network systems, can occur due to infrastructure failures, while severe weather events can cause significant damage to infrastructure and potentially trigger a crisis. Acts of terrorism or militia targeting critical infrastructure pose serious challenges to energy consumers as well. Political events such as regime changes can cause disequilibrium in the energy system, including disruption in oil and gas supply value chains, resulting in shortages. Furthermore, fuel shortages can occur due to excessive and wasteful use of fuels.

Although an authoritative definition of the concept of energy shocks is lacking, this phenomenon, which has emerged as a prominent concern since the 1970s, can be broadly comprehended as an occurrence within the energy system. This occurrence entails a substantial constriction in the availability of energy resources to an economy, resulting in a noteworthy bottleneck [47]. One notable demand–supply interaction feature of energy shock, particularly in the short-term – is price hikes or inflation during supply shock in the energy sector – which can simply be reasoned from the theoretical relationship between scarcity and price in microeconomics. A dominant reference period in the literature is the energy crisis of the 1970s, during which major industrial countries faced substantial petroleum shortages. This crisis had far-reaching effects, including a stock market crash, rampant inflation, and high unemployment rates. The two most severe crises during this period were the 1973 oil crisis and the 1979 energy crisis. While contestable, some commentators believe the energy shocks, especially from 1973 to 1979, significantly increased and sustained the sovereign debt burden of developing countries, as poor-oil-importing countries such as Ghana were compelled to borrow from commercial banks and economies with high-interest rates due to inflation tightening policy.[7]

As a commonplace is that the 1973 oil crisis occurred when Arab members of OPEC imposed an embargo against the United States and other countries supporting Israel during the Arab-Israeli War. The embargo prohibited petroleum exports to these nations and led to cuts in oil production. Regarding mechanism analysis, the 1979 energy crisis was triggered by decreased oil output following the Iranian Revolution. The outbreak of the Iran-Iraq War further disrupted oil production in the major Persian Gulf nations, leading to economic recessions in various oil-dependent. Oil prices did not return to pre-crisis levels until the mid-1980s.

The 1990 oil price shock occurred in response to the Iraqi invasion of Kuwait. The invasion and subsequent embargo caused a significant shift in the oil market. Concerns initially focused on the loss of crude oil supplies, and the price spike, lasting nine months, contributed to the recession of the early 1990s. Although less severe than the crises of the 1970s and 1980s, it still had a notable impact, especially among the industrialized club economies.

At this juncture, it is pertinent to highlight that the underlying factors behind energy crises, particularly in relation to crude products, exhibit a dynamic nature. The intricacies associated with these factors have shown a consistent inclination toward escalation as we approach the new millennium. Notably, during the 2000s, the ramifications stemming from oil shocks deviated from those witnessed in preceding decades. For example, in the writing period, the predominant factor behind the escalation in energy prices was the disequilibrium within the fundamental dynamics of the crude oil markets. This imbalance is juxtaposed with a constrained expansion of supply and accentuated by the COVID-19 pandemic, Russia-Ukraine tensions, and general speculative activities within global markets.

Table 2 shows crude oil prices from 2002 to 2022. To mitigate any possible bias regarding the COVID-19 pandemic and the Russia-Ukraine war in our analysis, 2019

---

[7] See for example, https://www.imf.org/external/about/histdebt.htm. Seen 30 July, 2023.

Year	Dubai ($/bbl)	Brent ($/bbl)	Nigerian Forcados ($/bbl)	West Texas Intermediate ($/bbl)
2002	23.60	25.02	25.04	26.16
2003	26.75	28.83	28.68	31.06
2004	33.51	38.27	38.13	41.49
2005	46.78	54.52	55.69	56.59
2006	61.48	65.14	67.07	66.04
2007	67.92	72.39	74.48	72.20
2008	94.28	97.26	101.43	100.06
2009	61.14	61.67	63.35	61.92
2010	77.78	79.50	81.05	79.45
2011	105.93	111.26	113.65	95.04
2012	109.06	111.67	114.21	94.13
2013	105.47	108.66	111.95	97.99
2014	97.02	98.95	101.35	93.28
2015	51.22	52.39	54.41	48.71
2016	41.02	43.73	44.54	43.34
2017	53.02	54.19	54.31	50.79
2018	70.15	71.31	72.47	65.20
2019	63.71	64.21	64.95	57.03
2020	42.41	41.84	42.31	39.25
2021	68.91	70.91	69.76	68.10
2022	96.38	101.32	101.40	94.00

Source: Author's compilation based Statistical Review of World Energy (2023 72nd Edition).

Table 2.
Crude oil prices in US dollars per barrel (2002–2022).

was chosen as a reference year. A general observation is that, from 2019 to 2020, global annual average oil prices witnessed a downward trend, suggesting an immediate effect of the COVID-19 lockdown and border restriction measures to contain the pandemic on the oil markets. In particular, with Brent averaging around $101 per barrel, oil prices experienced a rise of approximately 42% by the end of the year. It increased from 64.21 (US $/bbl) in 2019 pre-COVID-19 to 101.32 (US $/bbl) in 2022. Nonetheless, examining the aforementioned correlations carefully is crucial, considering both systematic and idiosyncratic risks specific to each country's circumstances. In certain nations, the observed impacts primarily underscore underlying structural vulnerabilities within their energy sub-sector. For instance, consider Ghana, where the escalation of energy costs – particularly regarding fossil fuels – is partly attributed to substantial debts in the energy sector, instances of corruption and mismanagement, inefficiencies in transmission and distribution, as well as the transition toward electricity generation through gas-fired thermal plants.

Overall, energy crises have historically stemmed from a range of factors, including shortages, conflicts, market manipulations, government interventions, infrastructure failures, geopolitical events, and excessive consumption. These crises have had significant economic and social repercussions, shaping the energy landscape and highlighting the importance of energy security.

7. Corruption and energy security

Corruption, generally, is seen as a drag on society's progress – economically, socially, politically, ecologically, or even, if verifiable, spiritually. For a straightforward discussion, I resort to a dictionary definition of "corruption" and its relationship with energy security. According to the Collins COBUILD Dictionary of CD-Rom (2006), the notion of corruption can be defined as "is dishonesty and illegal behavior by people in positions of authority or power". Despite the emphasis placed by the 2030 Sustainable Development Goals and international development initiatives on the pivotal role of energy access in socioeconomic advancement, a paucity of comprehension persists regarding the degree to which corruption within energy systems by prominent entities can either exacerbate or alleviate security concerns. Discussions at the beginning of the text accentuated that energy security ensures a stable energy supply crucial for development. However, corruption can threaten energy security goals by distorting market deficiencies, discouraging investments, and threatening the profitability, equity, and resilience of energy infrastructure network systems. As reported in [48], a recent study on Latin American societal elites reveals that hindrances to expanding wind and solar energy capacity and establishing larger cross-regional power systems primarily include corruption, bureaucratic challenges, insufficient global coordination, and substantial public-private investments. This section explores a limited mechanism through which energy corruption can undermine energy security.

Corruption, either from country to country, person to person, institution to institution, perceived or real, can significantly distort the energy markets and the governing institutions. Corruption *can* skew energy markets by favoring well-connected entities, reducing competition, and disrupting supply reliability due to subpar infrastructure. A case in point is that energy systems involve mega infrastructure investment, with millions of dollars in the North and South. Another lens of thinking about the energy-in(security)-nexus is that pervasive energy sector corruption can possibly result in distorted investment priorities. Corrupt practices often divert funds meant for energy infrastructure development and maintenance into the pockets of individuals or groups. This diversion can result in inadequate investment in critical energy projects, such as power plants, pipelines, and grids, leading to insufficient energy supply and frequent blackouts. For instance, if funds intended for upgrading power generation facilities are embezzled by corrupt officials, the energy system's capacity may remain insufficient to meet demand.

Arguably, irrespective of the political systems of a country – market or centrally planned economy, corruption in the energy sector can be a fertile ground for inefficient resource allocation. From a standpoint rooted in either political economy or political ecology, it can be duly posited that a fundamental imperative of a state resides in the judicious utilization of its collective resources, encompassing vital elements such as energy, to optimize the overall welfare of its populace, thereby effecting the reduction or mitigation of adverse welfare circumstances. Linked to the above, the United Nations' view on energy services and development, as indicated in the Seventh Goal of the SDGs (SDG 7), is that "The survival and advancement of humanity hinge upon the availability of energy for the purpose of household heating, the production of commodities, and the facilitation of communication across extended distances". This places energy resources at the center of multidimensional development thinking, including, ecological, social, institutional, economic, and political well-being. However, corruption can lead to the misallocation of energy resources, favoring politically connected parties or personal interests instead of optimizing resource utilization. For example, oil exploration and mining licenses might be awarded to companies based on bribes rather than their technical expertise, leading to ineffective resource management and decreased energy output [49, 50]. Furthermore, sector corruption can result in the erosion of regulatory frameworks. While not a novel idea, it is worth mentioning in the unique context of energy security discussions that corruption can weaken regulatory institutions responsible for overseeing the energy sector, leading to reduced accountability and enforcement of rules. Regulatory bodies that are compromised by corruption might overlook safety standards, environmental regulations, and fair market practices, ultimately risking energy supply disruptions and environmental degradation. Again, where it goes unchecked, energy sector corruption, especially mega oil, gas, electricity, nuclear plants, etc., can lead to unjustifiable higher costs and reduced competitiveness. Corruption often introduces inefficiencies and unnecessary costs into the energy value chain. When bribes and kickbacks are involved in procurement processes, costs can skyrocket, affecting the 4 A's of energy security for consumers and businesses. This reduced cost-effectiveness can make the country's industries less competitive globally. In addition, the cancer of energy corruption can repel foreign investment and energy-related foreign financial aid. Yes! Corruption in the energy sector can deter foreign investment! Investors (except corrupt ones) hesitate to commit capital to a country with an unreliable and unpredictable energy supply due to corruption, as it increases the perceived business risks. This lack of investment can hinder the development of new energy projects and technologies, further exacerbating energy security concerns. As an illustration, within the realm of contemporary indices utilized by organizations and foreign investors to conduct country risk analyses, the corruption index is a pivotal determinant in the decision-making process [51, 52].

Energy sector corruption may severely impact international environmental treaties and conventions, such as the Paris Agreement of 2015, intended to cut anthropogenic greenhouse gas emissions in the energy sector. In that, corruption can obstruct renewable energy development efforts. In many cases, corruption-driven policies can favor traditional energy sources over renewable alternatives. For instance, bribes and kickbacks from fossil fuel interests might lead to delayed adoption of renewable energy projects, hindering the transition to cleaner and more sustainable energy sources. Lastly, corruption, in general, can lead to social unrest and political instability: Energy shortages caused by corruption can lead to public dissatisfaction and unrest, as citizens suffer from blackouts, fuel shortages, and increased energy costs. This dissatisfaction can escalate into broader political instability, potentially affecting the country's stability and governance.

The detrimental impact of corruption within the energy sector on energy security is evident and far-reaching. Corrupt practices, such as bribery, embezzlement, and nepotism, weaken institutional frameworks, hinder efficient resource allocation, and undermine investor confidence. These actions lead to reduced investment in critical energy infrastructure, inefficient resource utilization, and distorted market competition. Consequently, energy supply disruptions, volatile prices, and unreliable access to energy resources become prevalent, jeopardizing a nation's energy security.

8. Demographic transition and energy security

For this text, we adopt the definition of demographic transition provided by Tulchinsky and Varavikova [53]. According to these authors, population transition refers to "a long-term trend of declining birth and death rates, resulting in substantive change in the age distribution of a population" ([53], p. 103). Demographic change can have significant implications for energy security. For instance, in sub-Saharan Africa, the large size of the population without access to clean cooking fuels and electricity is often attributed to the fact that the rate of population growth is higher than the annual percentage increase in electricity and clean cooking fuels and technology by organizations such as the International Energy Agency. However, [53] have provided five stages of population transition: namely, *Traditional, Transitional, Low stationary, Graying the population, and regression stage*. These stages, respectively, follow high and balanced birth and death rates, falling death rates and sustained birth rates, low and balanced birth and death rates, an increased proportion of older people as a result of decreasing birth and death, and low birth rates, migration, or rising death rates among young adults, respectively, according to [53]. Our world is typically characterized by the aforementioned stylized demographic transitions. It is therefore important to review the concept energy security from the above reality. Here are some ways in which demographic change can impact energy security.

First, is population growth. As commonly mentioned in the literature, especially economic development theories perspectives, as populations increase, there is a greater demand for energy to meet the needs of households, businesses, and industries. Arguably, while the mid-twentieth century marked the completion of the demographic transition, the same period marked the beginning of a population spurt in Asia, Latin America, and Africa [54]. Rapid population growth can strain energy resources and infrastructure, leading to challenges in ensuring a reliable and secure energy supply [55]. Secondly, and related to the above, is urbanization and energy security nexus. The process of urbanization, where people move from rural areas to cities, can result in concentrated energy demands. Urban areas tend to have higher energy consumption per capita than rural areas due to increased access to modern amenities, transportation needs, and higher population densities. Meeting the energy needs of rapidly growing cities requires robust and resilient energy systems to maintain energy security. Thirdly, demographic transition changes energy consumption patterns: Energy consumption is integral to social practices and social change [56, 57]. Thus, demographic changes, such as shifts in lifestyle, income levels, and cultural preferences, can influence energy consumption patterns. For example, as standards of living improve, there is often an increase in energy-intensive activities such as air conditioning, heating, and the use of electronic devices. These changing consumption patterns can place additional stress on energy systems, particularly if they are not adequately planned and managed.

As mentioned, *graying of the population* is one of the stages of population transition – or simply "aging population". Based on the population pyramid, an aging population constitutes an increase in the over 65 years old people. Recent estimates show that the population aged 65 years and above will double in the next three decades, reaching about 1.6 billion by 2015. Hong Kong, Japan, and South Korea are expected to have the highest population above 65 years. As populations in many countries continues to age, it can have implications for energy security. Older adults may have specific energy needs for heating or cooling due to health concerns, and ensuring their energy needs are met becomes crucial, especially in severe winters and summers. Additionally, an aging workforce in the energy sector can pose challenges in terms of knowledge transfer and maintaining the necessary expertise to operate and maintain energy infrastructure effectively.

In the context of low-middle-income countries, the aging population in advanced economies directly or indirectly impacts SDG 7.a – which calls for international public financial flows to developing countries to support clean energy and technologies. The aging population means increased demand for fiscal allocations to support the elderly population. This can lead to substantial cuts in financial aid to energy-poor economies to undertake infrastructure expansions and research activities. For instance, At the time of writing, Tokyo has announced 3.5 trillion yen ($25 billion) to finance the country's pro-natal policy – policies intended to encourage more children or boost the fertility rate [58].

Policymakers and energy planners must consider these demographic factors when formulating energy policies and strategies. By understanding and addressing the specific energy needs and challenges associated with demographic change, countries can enhance their energy security and promote sustainable and inclusive energy systems.

9. Power systems harmonics and energy security

The term "power system harmonics", denoted as "voltage and current distortions" [59], is a prime focus for physical scientists, especially electrical engineers [59]. Many researchers argue that the phenomenon of power system harmonic is not new, especially in the field of electrical engineering [60, 61]. For instance, while Salam [60] suggests that back in 1916, the scientist Steinmetz extensively examined and documented the impact of harmonics on three-phase power systems [60], Santoso et al., [62] the issues of power system harmonic and power quality "has become one of the most prolific buzzwords in the power industry since the late 1980s" ([62], p. 4). While the literature on system power harmonic might have matured, its integration into the analysis of energy security is under represented. Since our perspective on energy security largely falls in the social science domain, we decided to align our discussions with scholars who describe power system harmonic as one of the *quality* issues facing electric power systems. This view is significant that the energy goal of the United Nations' Sustainable Development Goals (SDGs) (SDG 7) is dedicated to "*access to affordable, reliable, sustainable, and modern energy for all by 2030*". This means that beyond the question of expanding energy access through infrastructure networks, the quality of the energy services is equality, an energy security issue at both micro and macro levels. The literature agreed that there are two main types of power system quality problems: *voltage-related problems* (e.g., harmonic, swells, blackouts, sags, etc.) and *current-related problems* (e.g., current leakage, electromagnetic interference, etc.). It is probably within this context that [59] contextualized harmonic as "distortions in the form of *voltage* and *current*". If we put the above in the lens of the *reliability* target of SDG 7, it makes sense to see the same as a threat to energy security. However, the authors [59] have expressed concern that despite clear harmonic impacts on power and system reliability, harmonics from renewable energy sources (RES) are poorly understood, and methods to eliminate them are underdeveloped.

In advanced economies, the dependability (reliability) of electricity provision, gauged by the frequency and duration of power supply disruptions, presently stands at a notably elevated level [63]. Low-middle-income and emerging economies, however, face energy insecurity from the phenomenon of harmonic distortions. This is primarily due to a weak infrastructure network system. For example, the authors [64] explore the quality of electricity services using open-ended interviews in Unguja (Tanzania). These researchers find that "*Fluctuations result in dim lights at best and power outages and broken appliances at worst, denying many Unguja residents the expected benefits of access to modern energy*". If unresolved, power system harmonic may result in energy losses, reduce power quality, damage equipment and household appliances, interfere with communication network systems, and cause instability in renewable energy integration. To ensure a true resilience sustainable energy development, countries should develop their own power quality standard with measurable indicators to assess progress and project possible risks, therefore.

Amidst climate change and extreme weather events, power system harmonics can worsen power system challenges, causing heightened energy losses, equipment damage, and compromised power quality. These effects strain resources and hinder power restoration post-disruptions. Consequently, addressing harmonic concerns is pivotal for ensuring power system resilience and reliability amidst shifting climate patterns and severe weather occurrences.

10. Energy security in climate change and uncertainty regimes

So, how should we think about energy security in the era of Anthropocene that is marked by climate change and variability (extreme weather events), pandemics, geopolitical tensions in energy resource-rich countries, divergence demographic transition between the global South and North, high unemployment low-income economies, natural and human-induced disasters, cyberattack of critical energy infrastructure, terrorism, poverty in its multidimensionality and uncertainties? As this chapter demonstrates, energy security, traditionally defined as the assured availability of energy resources at affordable prices, has long been a paramount concern for nations and global entities. Academic literature on energy security is reasonably rich, with multiple conceptual and operational definitions [65]. However, in an era characterized by climate change, resource depletion, high energy inflation, fuel price fluctuation, and geopolitical uncertainties, the concept of energy security demands a profound reevaluation. The transition toward resilient and sustainable energy development emerges as a compelling paradigm shift, representing an intricate interplay between environmental, economic, and social dimensions. This chapter expounds upon the evolution of energy security into a resilience-based framework, emphasizing the imperative of sustainable energy practices. Toward the concluding part of this chapter, we should think about the *science* of the regime: that climate change amplifies the occurrence, severity, and erratic nature of perturbations and pressures, thereby bestowing notably adverse repercussions upon societies exceedingly reliant on natural endowments. Simultaneously, vulnerability is also on the rise, accompanied by an escalation in the multifaceted and intricate nature of violence. It is projected that by the year 2030, when the United Nations Agenda for Sustainable Development elapses, in excess of 60% of the global impoverished populace will reside within contexts characterized by fragility [66]. Although social protection has emerged as a consequential policy domain within numerous low- and middle-income countries that appears as interventions against the above growing uncertainties, a substantial portion of the global populace, accounting for about 55%, remains devoid of any social protection benefits. As the pre-COVID-19 pandemic and Russia–Ukraine war, this deficiency is highly pronounced in sub-Saharan Africa, where 87% of individuals lack coverage, and in Asia and the Pacific, where the corresponding proportion is estimated at 61% ([67], p. 1). An energy security worry is that these regions constitute more than 85% of the population without access to modern energy services [68]. Therefore, we should think about integrating the imperative of resilience in energy security thinking.

10.1 Rethinking energy security as resilience sustainable energy development

10.1.1 From security to resilience

The historical notion of energy security was entrenched in the assurance of a steady supply of conventional fossil fuels, often tied to geopolitical stability, as established by relevant literature in the text. This approach, while providing short-term stability, remains ecologically and economically unsustainable. With the realization of the finite nature of fossil fuels, energy poverty and vulnerability, and the escalating consequences of climate change, energy security must transcend the traditional definition. It must now encapsulate not only the popular *4 A's – affordability, accessibility, acceptability, and availability* [17], but also the capacity to adapt and rebound in the face of disruptions and shocks. While challenging, especially in the unique context of low-middle-income economies with inequitable energy and weak infrastructure network systems, it should integrate a framework of thinking about energy security and resilience as mutually reinforcing and interlinking variables. For instance, while energy security is largely a component of national security strategies, resilience communicates "the property of the energy systems". According to [69], resilience refers to the ability of women and men to realize their rights and improve their well-being despite shocks, stresses, and uncertainty.

10.1.2 Resilience: The new energy paradigm

Resilience, defined as the ability of a system to withstand disturbances and recover its equilibrium, offers a comprehensive framework for rethinking energy security. A resilient energy system recognizes the multifaceted interdependencies among energy sources, infrastructure, ecosystems, and societal needs. It acknowledges the intricate web connecting energy, environment, and economy. By diversifying energy sources, incorporating distributed generation, and emphasizing local production, resilience-based energy systems reduce vulnerabilities to supply disruptions while fostering innovation and fostering localized economic growth. Against this logic, political economy compels the state to bear the duty of securing energy for all in a manner that is compatible with the health of the social-ecological systems' health. Resilient sustainable energy development thinking questions policymakers the extent to which secured energy services for population can be sustained in the event of shocks to ecological, social, and economic systems – using fresh experiences from the COVID-19 and Russia-Ukraine geopolitical tensions.

10.1.3 Balancing act: Policy and innovation

The journey toward resilience-based sustainable energy development necessitates a strategic confluence of policy innovation and technological advancement. Governments and international bodies must foster regulatory frameworks that incentivize the adoption of renewable energy sources, promote energy efficiency, and encourage investment in research and development. Innovations in energy storage, smart grids, and demand-side management are pivotal in enhancing energy system flexibility and adaptability.

10.1.4 Sustainability: The bedrock of resilience

Central to the transition from energy security to resilience-based sustainable energy development is the integration of sustainability principles. Sustainable energy practices minimize environmental degradation, promote efficient resource utilization, and mitigate greenhouse gas emissions. Renewable energy technologies, such as solar, wind, hydro, and geothermal, offer not only a dependable source of power but also a means to mitigate the adverse impacts of climate change. These technologies align with the Paris Agreement's broader objectives, enhancing energy security and global environmental well-being.

10.1.5 Conclusion

In conclusion, the evolution of energy security from a narrow focus on supply reliability to a broader resilience-based approach represents a paradigm shift reflective of the contemporary global landscape. The imperative of sustainable energy development demands that nations and institutions reevaluate energy security through the lens of environmental stewardship, societal well-being, and economic robustness. By embracing renewable energy sources, enhancing energy efficiency, and bolstering technological innovation, the international community can forge a more secure, resilient, and sustainable energy future. This transition, while undoubtedly complex, holds the promise of not only safeguarding energy access but also fostering a harmonious coexistence between humanity and the planet.

Conflict of interest

The authors declare no conflicts of interest related to this academic paper.

Author details

Smart Edward Amanfo[1*] and Joseph John Puthenkalam[2]

1 Independent Researcher, Seiha English Academy, Fukuoka, Japan

2 Graduate School of Global Environmental Studies at Sophia University, Tokyo, Japan

*Address all correspondence to: smart6783@gmail.com

References

[1] Smil V. Energy and Civilization: A History. London, England: MIT press; 2017

[2] Radovanović M. Chapter 8 - Energy security. In: Radovanović M, editor. Sustainable Energy Management. 2nd ed. Boston: Academic Press; 2023. pp. 279-304

[3] Chester L. Conceptualising energy security and making explicit its polysemic nature. Energy Policy. 2010; **38**(2):887-895

[4] Sovacool BK, Brown MA. Competing dimensions of energy security: An international perspective. Annual Review of Environment and Resources. 2010;**35**(1):77-108

[5] Van de Graaf T, Zelli F. Actors, Institutions and Frames in Global Energy Politics. London: Palgrave Macmillan UK; 2016. pp. 47-71

[6] Novikau A. Energy Security: Evolution of a Concept. Cham: Springer International Publishing; 2020. pp. 1-4

[7] Lubell H. Security of supply and energy policy in Western Europe. World Politics. 1961;**13**(3):400-422

[8] Kohl WL. National security and energy. In: Encyclopedia of Energy. Amsterdam, Netherlands: Elsevier; 2004. pp. 193-206

[9] Zhao H. Energy security. In: The Economics and Politics of China's Energy Security Transition. United Kingdom: Elsevier; 2019. pp. 99-120

[10] Miller LB. Energy, security and foreign policy: A review essay. International Security. 1977;**1**(4): 111-123

[11] Willrich M. International energy issues and options. Annual Review of Energy. 1976;**1**(1):743-772

[12] Bajracharya TR, Shakya SR, Sharma A. Dynamics of energy security and its implications. In: Handbook of Energy and Environmental Security. United Kingdom: Elsevier; 2022. pp. 13-25

[13] Azzuni A, Breyer C. Definitions and dimensions of energy security: A literature review. WIREs Energy and Environment. 2018;**7**(1):e268

[14] Jiang Y, Liu X. A bibliometric analysis and disruptive innovation evaluation for the field of energy security. Sustainability. 2023;**15**(2):969

[15] Sovacool BK, Valentine SV, Bambawale MJ, Brown MA, de Fátima Cardoso T, Nurbek S, et al. Exploring propositions about perceptions of energy security: An international survey. Environmental Science and Policy. 2012; **16**:44-64

[16] Sovacool BK. Differing cultures of energy security: An international comparison of public perceptions. Renewable and Sustainable Energy Reviews. 2016;**55**:811-822

[17] Cherp A, Jewell J. The concept of energy security: Beyond the four As. Energy Policy. 2014;**75**:415-421

[18] Fares A, Habibi H, Awal R. Extreme events and climate change: A multidisciplinary approach. In: Climate Change and Extreme Events. New Jersey: Elsevier; 2021. pp. 1-7

[19] Stockemer D. The nuts and bolts of empirical social science. In: Quantitative Methods for the Social Sciences: A Practical Introduction with Examples in

SPSS and Stata. Switzaland: Springer International Publishing; 2018. pp. 5-22

[20] David A, Nagle G. Environmental Systems and Societies. United Kingdom: Oxford Univerisity Press; 2020

[21] Tosam MJ, Ambe JR, Chi PC. Global Emerging Pathogens, Poverty and Vulnerability: An Ethical Analysis. Cham: Springer International Publishing; 2019. pp. 243-253

[22] Overland I, Juraev J, Vakulchuk R. Are renewable energy sources more evenly distributed than fossil fuels? Renewable Energy. 2022;**200**:379-386

[23] Yao X, Yasmeen R, Padda IUH, Shah WUH, Kamal MA. Inequalities by energy sources: An assessment of environmental quality. PLoS One. 2020;**15**(3):e0230503

[24] Kisel E, Hamburg A, Härm M, Leppiman A, Ots M. Concept for energy security matrix. Energy Policy. 2016; **95**:1-9

[25] Leal-Arcas R, Grasso C, Ríos JA. Chapter 6: Diversifying EU Energy Supply to Improve EU Energy Security. Cheltenham, UK: Edward Elgar Publishing; 2016

[26] Streimikiene D, Siksnelyte-Butkiene I, Lekavicius V. Energy diversification and security in the EU: Comparative assessment in different EU regions. Economies. 2023;**11**(3):83

[27] Intharak N, Julay JH, Nakanishi S, Matsumoto T, Sahid E, Ormeno AAG, et al. A quest for energy security in the 21st century: Resources and constraints. In: Technical Report. Japan: Asia Pacific Energy Research Centre; 2007

[28] Knodt M, Kemmerzell J. Energy Governance in Europe: Introduction. Cham: Springer International Publishing; 2022. pp. 3-16

[29] Amirova-Mammadova S. New Geopolitics of the Southern Gas Corridor. Wiesbaden: Springer Fachmedien Wiesbaden; 2018. pp. 159-190

[30] Global Energy Trends – 2023 Edition. Technical Report, Enerdata. 2023. Available https://www.enerdata.net/publications/reports-presentations/world-energy-trends.html [Accessed: August 06, 2023]

[31] Dunn C, Hess T. The United States is now the largest global crude oil producer [Online]. Energy Information Administration (EIA). 2018. [Accessed: July 22, 2023]

[32] Secretariat EC. International energy security: Common concept for energy producing, consuming and transit countries. Energy Charter Secretariat. Technical Report 6. 2015. pp. 1-28

[33] Japan's energy 2019: 10 questions for understanding ther current energy situation. In: Technical Report, Ministry of Economy, Trade and Industry. Japan: Agency for Natural Resources and Energy; 2020

[34] Hayashi M, Hughes L. The Fukushima nuclear accident and its effect on global energy security. Energy Policy. 2013;**59**:102-111

[35] IEA. apan 2021: ENergy Policy Review. 2021. Online. Available from: https://www.iea.org/reports/japan-2021 [Accessed: July 15, 2023]

[36] The Energy Institute. Statistical Review of World Energy. Online. 2023. Available from https://www.energyinst.org/__data/assets/pdf_file/0004/1055542/EI_Stat_Review_PDF_single_3.pdf [Accessed: July 20, 2023]

[37] BP. BP Statistical Review of World Energy. 2018. Online. Available at

https://www.bp.com/content/dam/bp/business-sites/en/global/corporate/pdfs/energy-economics/statistical-review/bp-stats-review-2018-full-report.pdf

[38] Bickford A, Wright JD. Militaries and militarization, anthropology of. In: International Encyclopedia of the Social & Behavioral Sciences. Second ed. Oxford: Elsevier; 2015. pp. 483-489

[39] Czarny RM. Energy security: Contemporary challenges. In: The Nordic Dimension of Energy Security. Cham: Springer International Publishing; 2020. pp. 1-22

[40] Keshk AM. NATO and the Gulf Countries: An Analysis of the Fifteen Year Strategic Partnership. Singapore: Springer; 2021

[41] NATO. Strategic Concept for the Defense and Scurity of the Members of the North Atlantic Treaty Organization. Online. Adopted by Heads of State and Governmentat the NATO Summit in Lisbon 19-20 November 2010. 2010

[42] Wang H, Xu Q, Wang H, Xu Q. Energy diplomacy of the major countries in the world. In: An Introduction to Energy Diplomacy: China's Perspective. Singapore: Springer Singapore; 2022. pp. 161-179

[43] Ritchie H, Roser M, and Rosado P. 2022. Energy. Our World in Data. Retrived from https://ourworldindata.org/energy

[44] EIA. World Oil Transit Chokepoints. 2017. Online. Available from https://www.eia.gov/international/content/analysis/special_topics/World_Oil_Transit_Chokepoints/wotc.pdf. July 25, 2023

[45] Lutmar C, Rubinovitz Z. Suez Canal Past Lessons and Future Challenges. Springer International Publishing AG; 2022

[46] Goldemberg J. Energy, environment and development. United Kingdom: Earthscan; 2010

[47] Kilian L. Energy price shocks. In: The New Palgrave Dictionary of Economics. London: Palgrave Macmillan UK; 2016. pp. 1-12

[48] Noor R, Sanda SN. Harnessing Win-Win Energy Geopolitics and Competitive Global Energy Market by Integrating Energy Efficiency. Singapore: Springer Nature Singapore; 2022. pp. 1-30

[49] Fhima F, Nouira R, Sekkat K. How does corruption affect sustainable development? A threshold non-linear analysis. Economic Analysis and Policy. 2023;**78**:505-523

[50] Fredriksson PG, Vollebergh HR, Dijkgraaf E. Corruption and energy efficiency in OECD countries: Theory and evidence. Journal of Environmental Economics and Management. 2004;**47** (2):207-231

[51] Bouchet MH, Fishkin CA, Goguel A, Bouchet MH, Fishkin CA, Goguel A. In search of early warning signals of country risk: Focusing on capital flight. In: Managing Country Risk in an Age of Globalization: A Practical Guide to Overcoming Challenges in a Complex World. Cham: Springer International Publishing; 2018a. pp. 413-442

[52] Bouchet MH, Fishkin CA, Goguel A, Bouchet MH, Fishkin CA, Goguel A. The root causes and consequences of political risk: From bad governance to wealth and political power concentration and social instability. In: Managing Country Risk in an Age of Globalization: A Practical Guide to Overcoming Challenges in a Complex World. Cham: Springer International Publishing; 2018b. pp. 289-318

[53] Tulchinsky TH, Varavikova EA. Measuring, monitoring, and evaluating the health of a population. In: The New Public Health. California: Elsevier; 2014. pp. 91-147

[54] Ness GD. Population growth and energy. In: Encyclopedia of Energy. Amsterdam, Netherlands: Elsevier; 2004. pp. 107-116

[55] Nepal R, Paija N. Energy security, electricity, population and economic growth: The case of a developing south Asian resource-rich economy. Energy Policy. 2019;**132**:771-781

[56] Nature Journal's Editorial Board. The role of society in energy transitions. Nature Climate Change. 2016;**6**(6):539-539. Available from: https://www.nature.com/articles/nclimate3051

[57] Horta A. 31energy consumption as part of social practices: The alternative approach of practice theory. In: Davidson DJ, Gross M, editors. Oxford Handbook of Energy and Society. New York: Oxford University Press; 2018

[58] Trogen PC, Xu Y. Population policy, China. In: Farazmand A, editor. Global Encyclopedia of Public Administration, Public Policy, and Governance. Cham: Springer International Publishing; 2017. pp. 1-7

[59] Eroğlu H, Cuce E, Mert Cuce P, Gul F, Iskenderoğlu A. Harmonic problems in renewable and sustainable energy systems: A comprehensive review. Sustainable Energy Technologies and Assessments. 2021;**48**:101566

[60] Salam A. Power system harmonics. In: Fundamentals of Electrical Power Systems Analysis. Singapore: Springer Singapore; 2020. pp. 461-495

[61] Rosa FCDL. Harmonics, Power Systems, and Smart Grids. 2nd ed. Boca Raton: Taylor & Francis Group; 2015

[62] Santoso S, McGranaghan MF, Dugan RC, Beaty HW. Electrical Power Systems Quality. 2nd ed. New York: McGraw-Hill Education; 2012

[63] Perera S, Elphick S, Perera S, Elphick S. Chapter 4 - Impact and management of power system harmonics. In: Applied Power Quality: Analysis, Modelling, Design and Implementation of Power Quality Monitoring Systems. United Kingdom: Elsevier; 2023. pp. 71-130

[64] Jacome V, Klugman N, Wolfram C, Grunfeld B, Callaway D, Ray I. Power quality and modern energy for all. Proceedings of the National Academy of Sciences. 2019;**116**(33):16308-16313

[65] Novikau A. Conceptualizing and redefining energy security: A comprehensive review. In: China's Energy Security. Singapore: Publishing Co Pte Ltd; 2021. pp. 37-59

[66] Henly-Shepard S, Zommers Z, Levine E, Abrahams D. Climate-resilient development in fragile contexts. In: Resilience: The Science of Adaptation to Climate Change. Elsevier; 2018. pp. 279-290

[67] Loewe M, Schüring E. Chapter 1: Introduction to the Handbook on Social Protection Systems. Cheltenham, UK: Edward Elgar Publishing; 2021

[68] IEA. World Energy Outlook 2022. 2022. [Online]. License: CC BY 4.0 (report); CC BY NC SA 4.0

[69] Jeans H, Thomas S, Castillo G. The future is a choice: The Oxfam framework and guidance for resilient development. [Online] [Accessed: October 19, 2023]